Risk Management for Geotechnical Engineering

Risk Management for Geotechnical Engineering: Hazard, Risks and Consequences covers the application of risk management for soil and rock engineering projects, and the preparation of reliable designs that account for uncertainty.

This book discusses qualitative risk assessments based on experience and judgement, as well as quantitative risk analysis using probabilistic methods and decision analysis to optimize designs. Many examples are included of how risk management can be applied to geotechnical engineering, with case studies presented for debris flows, rock falls, tunnel stability, and dam foundations. This book also discusses the issues of liability insurance and contract law related to geotechnical engineering.

This comprehensive book is ideal for practicing geotechnical engineers, addressing the challenges of making decisions in circumstances where uncertainties exist in site conditions, material properties and analysis methods.

Duncan C. Wyllie is Principal at Wyllie & Norrish Rock Engineers, Vancouver, Canada. He has over 50 years of experience in applied rock engineering and his other books include *Rock Slope Engineering*, *Rock Fall Engineering*, and *Foundations on Rock*, also published by CRC Press.

Risk Management for Geotechnical Engineering
Hazard, Risks and Consequences

Duncan C. Wyllie Ph.D.

CRC Press is an imprint of the
Taylor & Francis Group, an **informa** business

Cover image: Duncan C. Wyllie

First edition published 2024
by CRC Press
6000 Broken Sound Parkway NW, Suite 300, Boca Raton, FL 33487-2742

and by CRC Press
4 Park Square, Milton Park, Abingdon, Oxon, OX14 4RN

CRC Press is an imprint of Taylor & Francis Group, LLC

ISBN: 978-1-032-22266-0 (hbk)
ISBN: 978-1-032-22267-7 (pbk)
ISBN: 978-1-003-27186-4 (ebk)

DOI: 10.1201/9781003271864

Typeset in Sabon
by codeMantra

Contents

List of Figures

List of Tables

List of Symbols

A	Area of sliding plane (m^2)
ALARP	As Low As Reasonably Possible
b	Distance of tension crack behind crest of slope (m)
B	pattern rock bolting (tunnel support)
C	Capacity (kN)
c	Cohesion (kPa)
CCA	cast concrete liner (tunnel support)
D	Demand, Displacement force (kN)
DBR	Dispute Review Board
ESR	Excavation Support Ratio (tunnel support design)
EV	Expected value = ((event cost) × (probability of occurrence))
FS	Factor of Safety
HCV	Highest Conceivable Value
LCV	Lowest Conceivable Value
M	Margin of safety (probability analysis)
N	Population
P	Probability of occurrence
PF	Probability of Failure
PLL	Potential loss of life
q	Bearing pressure (kPa)
Q	Tunnelling rock mass quality index; bearing force (kN)
R	Resistance force, (kN); radius of reliability ellipse (reliability based design)
R	Correlation matrix (reliability based design)
RQD	Rock Quality Designation (drill core)
RRS	Steel sets (tunnel support)
S	Unreinforced shotcrete (tunnel support)
sb	Spot bolts (tunnel support)
sfr	Steel fibre reinforced shotcrete (tunnel support)
SRF	Stress Reduction Factor (tunnel design)
U	Water force on sliding plane (kN)
V	Water force in tension crack (kN)
VOL, VOSL	Value of Life, Value of Statistical Life ($, €)

z	Depth of tension crack (z)
z_w	Depth of water in tension crack (m)
β	Reliability Index
ψ	Discontinuity dip angle, or slope angle (°)
ϕ	Friction angle (°)
φ	Resistance Factor (LFRD design)
μ	Mean (average) of data set
γ	Unit weight (kN/m^3); Load Factor (LRFD design)
η	Importance Factor (building)
σ	Standard deviation of data set; normal stress (kPa)
$\cdot$	Multiplication symbol

NOTE

The recommendations and procedures contained herein are intended as guidelines, and prior to their use in connection with any investigation, design, report, specification or construction procedure, they should be reviewed regarding the full circumstances of such use. Accordingly, although care has been taken in the preparation of this book, no liability for negligence or otherwise can be accepted by the author or the publisher.

Preface

This book on risk management for geotechnical engineering has been written specifically for practicing engineers who are frequently faced with the challenges of making decisions in circumstances where significant uncertainties exist. Examples of uncertain conditions include the extent of investigation programs, appropriate factors of safety (or probability of failure) to use in design and the quantity of support to install in an excavation. Hopefully, this book will help to provide systematic and quantitative methods of assessing the level of risk on projects and incorporating uncertainties into the work. This book is aimed to be a middle ground between descriptive books that discuss programs that can be used for risk management, and more mathematical books that provide details on the statistics of geotechnical engineering.

Chapters are included on insurance and legal matters related to geotechnical engineering because both are important in the practice of engineering. That is, it is unwise to operate an engineering business where high levels of uncertainty exist, without the protection of Professional Liability Insurance (Errors and Omissions) against claims of negligence. Similarly, it is valuable to have a sound legal basis for engineering services in terms of ensuring that contractual conditions are equitable and understanding the precedents and defence of legal claims.

This book includes many examples of how risk management can be applied to geotechnical engineering, including four typical case studies for debris flows, rock falls, tunnel stability and dam foundations. The examples quoted mostly relate to geological engineering because that is the writer's field of expertise. However, the risk management principles described in this book are applicable to all fields of geotechnical engineering.

Preparation of this book has benefited greatly from the assistance of several people, some of whom I have worked with for many years, both professionally, and on my previous books. First, I would like to thank Tony Moore and Aimee Wragg with my publisher Taylor & Francis who have supported this book and my previous books. Also, I must thank Tyler Southam for his review of the technical aspects of the work, Jeff McLellan and Gareth McDonnell who reviewed the insurance chapter and Karen Weslowski for her review of the legal chapter. Their work has been invaluable. Much of the artwork has been prepared by Sonia Skermer, and help with research and organization of the material has been provided by Glenda Gurtina and Mehera Salah. I appreciate the assistance of them all.

Finally, I must thank my wife Airlie for her support and patience with yet another multi-year publication endeavour.

Duncan C. Wyllie, Ph.D.
Vancouver, Canada
2023

Previous books by Duncan Wyllie

- *Foundations on Rock* (editions 1 and 2) – 1992 and 1999
- *Rock Slope Engineering*, based on work by Evert Hoek and John Bray (editions 4 and 5) – 2010 and 2018.
- *Rock Fall Engineering* (edition 1) – 2015.

Chapter 1

Uncertainty and risk in geotechnical engineering

1.1 UNCERTAINTY IN GEOTECHNICAL ENGINEERING

The practice of geotechnical engineering encompasses a wide diversity of projects including tunnels, foundations and slopes. These projects may encounter a range of materials from weak soils to strong rock, constructed in a variety of climatic conditions from desert, temperate to polar and be constructed in terrain that may be easily accessible to challenging. For each of these conditions, uncertainty will often exist that needs to be incorporated into design. Examples of uncertainty include encountering zones of weak, saturated rock in tunnels, to unrecognized landslides that are activated by construction.

In order to successfully design, construct and operate geotechnical structures, it is beneficial to account for uncertainty in all aspects of the project, either quantitatively in which numerical values can be given to the uncertainty, or qualitatively in which experience and judgement are used to assess the degree of uncertainty. This information is then used to determine the project risk that is a function of probability of a hazard occurrence and its consequence (Morgenstern, 2018; Phoon & Ching, 2015; Hoek, 1994).

The issues of uncertainty and preparation of reliable geotechnical design have been encapsulated as follows (Morgenstern, 2000):

> The assurance of geotechnical performance would be enhanced if geotechnical engineering shifted from the promise of certainty, to analysis of uncertainty.

Furthermore, the vulnerability of people working in this field is clearly demonstrated by the situation in Italy where six scientists were fined and jailed for failing to predict the magnitude 6.3 earthquake that destroyed the town of L'Aquila in 2009.

DOI: 10.1201/9781003271864-1

Once the risk level has been determined, it can be compared to generally accepted societal risks to determine if the risk is acceptable, or whether remedial work is required. Examples of remedial work are slope stabilization to reduce the probability of landslide occurrence, or moving vulnerable structures to reduce the consequence of a landslide. Following the remedial work, reassessment of the risk would be required to make sure that the revised risk is acceptable.

This book discusses techniques to assess risk levels and determine whether they are acceptable. These techniques are well developed in engineering practice such that it is possible to apply a generally consistent approach to risk management for different projects.

In addition to the engineering aspects of risk management, this book discusses liability insurance coverage for engineering companies involved in project design and construction work, as well as legal issues related to claims and contracts. Insurance coverage and legal advice are essential components of most geotechnical projects because of the significant uncertainty in design and construction conditions, and the possibility of claims being made against engineering companies (Ferguson, 2021).

1.2 DESCRIPTION OF TERMS DEFINING UNCERTAINTY

Because the primary topic of this book is uncertainty, it is useful at the start to list various terms that are used to describe and/or define uncertainty. These terms each have a meaning related to the type of analysis that is being used, and the terms cannot be used interchangeably.

The following is a description of seven terms defining uncertainty, and the main sections in the book where they are addressed in more detail:

a. **Likelihood of event occurrence** – if the chance of an event occurring in the future has to be estimated based on relevant experience and expertise because no specific information on past occurrences is available, then likelihood of occurrence can be expressed by terms such as “possible” or “remote”. The term likelihood may be used in the early stages of a project when little site information is available.
b. **Annual probability of occurrence** – if a significant quantity of information is available of previous events, such a slope instability on a transportation system, then annual (temporal) probabilities can be calculated that can be used to predict future occurrences and plan appropriate stabilization work (see Section 2.4.2).

c. **Factor of safety (FS)** – the commonly used factor of safety can be expressed as the ratio: (FS=capacity/demand). Typical FS values may range from about 1.2 for low-risk, temporary structures to 3.0 for high-risk structures where the level of uncertainty in the design parameters is significant (see Section 2.1.2).
d. **Probability of failure (PF)** – the difference between the capacity and demand is termed the margin of safety, which is positive for a stable structure and negative for an unstable structure. If the capacity and demand are both expressed as probability distributions, then the margin of safety will also be a probability distribution. By convention, the negative portion of the margin of safety distribution is termed the "probability of failure (PF)". However, it is important to recognize that PF has no time component and, in reality, is an alternative expression for the factor of safety (see Section 2.4).
e. **Ultimate limit state (ULS)** – the ULS accounts for uncertainty in the design parameters by multiplying capacity parameters such as shear strength by partial factors that are less than 1.0 and multiplying demand parameter such as loads by partial factors that are great 1.0. The magnitude of the partial factors is selected such that the failure of the structure is highly unlikely (see Section 2.6).
f. **Serviceability limit state (SLS)** – the SLS uses the same analysis method as the ULS except that partial factors are closer to 1.0 such that elastic movement or settlement of the structure is possible. Effectively, the SLS has a lower factor of safety than the ULS (see Section 2.6).
g. **Likelihood in Bayesian analysis (L_z)** – Bayesian analysis involves updating an existing set of data, the values of which are defined by a Prior probability distribution. By making additional measurements of this parameter defining the Likelihood function and then using the Bayesian equation to update the Prior distribution and calculate a Posterior distribution of all the data (see Section 4.6). The Likelihood function, as used in Bayesian analysis, concerns the hypothesis relating to the prior and posterior distributions and is an important statistical function that is not related to the likelihood of an event occurrence as described in a) above.

1.3 RISK MANAGEMENT STRUCTURE – PROBABILITY AND CONSEQUENCE OF HAZARD OCCURRENCE

A typical structure for a risk management analysis, comprising nine well-defined tasks, is shown in Figure 1.1. The basis for this risk procedure is to first identify the risks (*Task 1*) and then analyze the data in terms of the likelihood or probability of its occurrence, and the consequence(s) that may develop if the event occurs (*Task 2*). The chance of an event occurring can

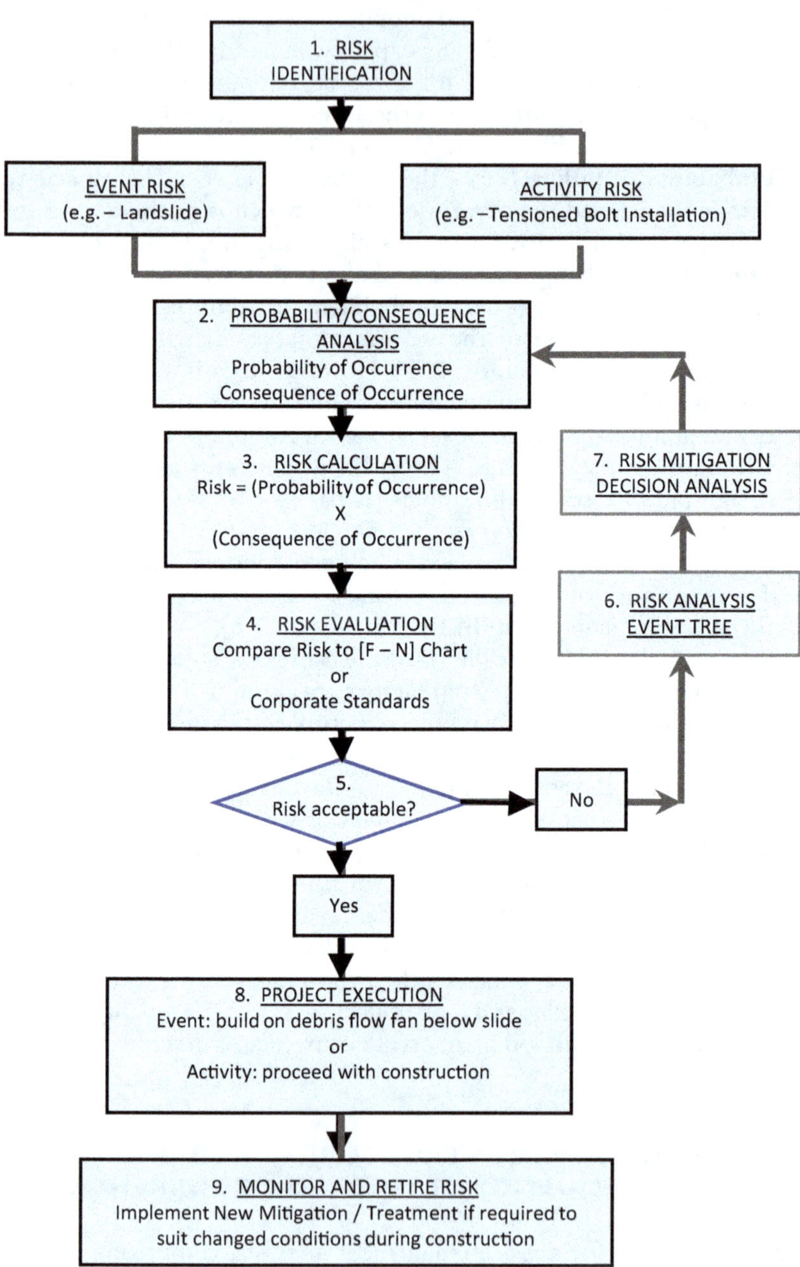

Figure 1.1 Structure of risk management programs.

be expressed as either "likelihood" if little information on past events is available and judgement must be used to assess its occurrence, or "probability" if information on past events is available and the probability of future occurrence can be calculated.

Using this information, risk is then defined in *Task 3* by the relationship:

$$Risk = (probability\ of\ occurrence) \times (consequence\ of\ occurrence) \quad (1.1)$$

The calculated risk can then be compared to levels of risk that are acceptable to society or to established corporate acceptable risk levels (*Task 4*). If the calculated risk is not acceptable (*Task 5*), then mitigation or treatments can be undertaken to reduce the probability of occurrence and/or the consequences such that the risk is reduced to an acceptable level (*Tasks 6* and *7*). This management procedure can be either qualitative if judgement or experience is required to estimate probabilities of occurrence, or quantitative if data are available on the frequency and magnitude of previous events and their consequences.

Once the risk is acceptable and the project is initiated, the risks identified in the planning and design phase of the project can be monitored and updated (*Task 8*). As the project develops, the risks may be retired if steps have been undertaken to eliminate or control the risk, or the risk may be carried forward into the operational stage of the project so that changing risk conditions can be addressed, as required (*Task 9*).

These tasks are addressed in the following chapters of the book:

- **Chapter 3** – Task 1
- **Chapter 4** – Task 2
- **Chapter 5** – Tasks 3, 4, 5
- **Chapter 6** – Task 6
- **Chapter 7** – Task 7

1.4 APPLICATION OF QUANTITATIVE RISK ANALYSIS

One of the most comprehensive applications of quantitative risk analysis is in Hong Kong where a combination of steep slopes, weathered rock, heavy rainfall events and high population density has resulted in the occurrence of landslides that have sometimes caused multiple fatalities (Ho & Ko, 2007). As a result of very serious consequences of these events, and recognition of continuing potentially hazardous slopes above residential buildings and highways, the Hong Kong Geotechnical Engineering Office has developed a Quantitative Risk Assessment (QRA) procedure that is applied to soil and rock cut slopes, fill slopes and retaining walls (Wong, 2005). The QRA system comprises a numeric rating system for slopes that includes an Instability

Score and a Consequence Score, with the Total Score being the product of these two scores.

This quantitative system has identified and ranked the most hazardous sites and allowed a focused stabilization plan to be developed. The separate scores for instability and consequence allow the hazards for each of these conditions to be clearly identified so that appropriate mitigation work can be planned. An important feature of this scoring system is that a potentially high-consequence event, but with very low probability of occurrence because remedial work has been carried out, can be clearly identified as a low-risk site in the slope database.

Similar risk management has been implemented for transportation systems such as a slope stability program on about 1,600 kilometres of track for the Canadian Pacific where each slope is assigned an Action Rating for required remedial work (Mackay, 1997), and for highways using a numeric scoring system to quantify risk (Wyllie, 1987; Pierson et al., 1990).

1.4.1 Types of risk – event risk and activity risk

Risk identification in *Task 1* examines two classes of risk, namely, event risk and activity risk, which are defined as follows.

- ***Events*** are typically landslides, debris flows and rock falls that have multiple occurrences but are mostly unpredictable, except perhaps that instability is often related to weather conditions such as heavy rainfall and freeze/thaw cycles, or to seismic ground shaking. The annual frequency of these events can be determined from past records where available or by the judgement of suitably qualified professionals based on their experience and interpretation of site conditions. These events can have consequences such as interruptions to traffic, damage to property or loss of life. Site data can be used to calculate annual risks that can be compared to acceptable risks for these conditions, and if the risk is unacceptable, appropriate mitigation or treatment can be implemented.
- ***Activities***, in contrast to events, activities are single occurrences of construction projects, such as excavation of soil at a stable slope angle, rock excavation that requires blasting, or installation of tensioned rock anchors to stabilize a slope or foundation. Possible uncertainties related to the activity are the experience of the construction crew, influence of weather on the work and unexpected geological conditions such weak rock. Potential consequences arising from these uncertainties are schedule delays and cost overruns. Analysis of the risks will identify the most likely causes of construction delays and cost overruns so that appropriate mitigation measures, such as verifying the contractor's experience and equipment, can be implemented before construction starts. Risk analysis may also show that additional investigation work may be beneficial to reduce uncertainty in specific aspects of the work.

Table 1.1 Examples of likelihood descriptions for event and activity risks

Event risk *(Frequency of event)*		Activity risk *(Likelihood of occurrence)*	
Event occurs annually	Annual frequency >1	New, complex activity for which crew has no previous experience	Very high likelihood of cost overrun and schedule delay
Event occurs approximately every 10 years	Annual frequency 0.1–1	Limited training available prior to start of project	High likelihood of cost overrun and schedule delay
Event occurs approximately every 10 to 100 years	Annual frequency 0.01–0.1	Experienced personnel available to provide guidance	Moderate likelihood of cost overrun and schedule delay
Event occurs approximately every 100 to 1,000 years	Annual frequency 0.001–0.01	Advice and training from specialists available prior to and during project	Project possibly on time and within budget
Event occurs every 1,000 to 10,000 years	Annual frequency 0.001–0.0001	Routine activity by experienced crew	Project likely to be on time and within budget

For both event and activity risks, the definitions of likelihood and consequence can be customized to suit project circumstances. Examples of likelihood definitions are shown in Table 1.1.

The structured approach to risk management shown in Figure 1.1 can be applied to a wide range of project types, which means that risk management for different projects will be consistent. Furthermore, comparing project risks to generally acceptable societal risk will be beneficial in designing projects to a uniform risk standard (Duzgun & Lacasse, 2005).

1.4.2 Risk matrix – likelihood of hazard occurrence and consequence

Task 2 in Figure 1.1 shows that analysis of risk involves two tasks. First, to estimate the likelihood of the event occurrence, and second, to examine the possible consequences of the event. The likelihood of occurrence can range from very high to very low, and the consequence from extreme to minimal. The selected values are either qualitative, based on experience and judgement, or quantitative if specific information is available on the frequency of previous events for example, or the consequence can be expressed as a monetary value.

Task 3 in Figure 1.1 involves plotting the event likelihood and consequence on a risk matrix, where the likelihood of the event occurrence is plotted on the vertical scale, and the consequence is plotted on the horizontal scale (Figure 1.2). The use of standard risk matrices allows risk for different projects to be analyzed consistently.

An example of a risk matrix application is slope stability on a transportation system. Regarding the likelihood of slope instability, records of past

RISK MATRIX (Qualitative)		Consequence of negative outcome				
		Minimal consequence	Minor consequence	Moderate consequence	Major consequence	Extreme consequence
Likelihood of event occurrence	Very high likelihood					Very high risk FS ≈ 2 – 2.5
	High likelihood					
	Moderate likelihood					
	Low likelihood					
	Very low likelihood	Very low risk FS ≈ 1.3 - 1.5				

Figure 1.2 **Risk matrix showing qualitative descriptions for likelihood and consequence, and target ranges of factor of safety to use in design.**

events show that they occur every 2–3 years, and sometimes every year – this would be a very high likelihood of an event occurrence. Regarding the consequence of these events, the use of warning systems and catchment structures means that slides rarely reach the road so the events have minor or minimal consequences to operations. For this situation, the risk would plot on the top, left corner of the matrix, indicative of medium risk.

The matrix shown in Figure 1.2 is a graphic expression of risk management that is probably carried out intuitively by geotechnical engineers during project investigation and design. That is, if the likelihood of a hazard occurring is very low and the consequence of the event is minimal, then the risk would be low, and the corresponding factor of safety of the design might be in the range of 1.3–1.5. However, for a very high likelihood of occurrence with extreme consequences, the factor of safety may be in the range of 2–2.5.

Where information is available on the frequency of previous events, the annual probability of occurrence can be calculated, and it may also be possible to quantify the consequences of landslides in terms of property damage and injury to persons. For these conditions, the risk matrix may be quantified by assigning values to each category of likelihood and each category of consequence from which a numerical risk can be calculated from the product of the likelihood and consequence scores. Examples of quantitative risk matrices are discussed in Chapter 3.

1.4.3 Acceptable risk

Task 4 in Figure 1.1 involves comparing the calculated risk with a commonly used measure of acceptable risk such as the [F – N] chart shown in Figure 1.3 where the frequency of an event occurrence [F] can be related

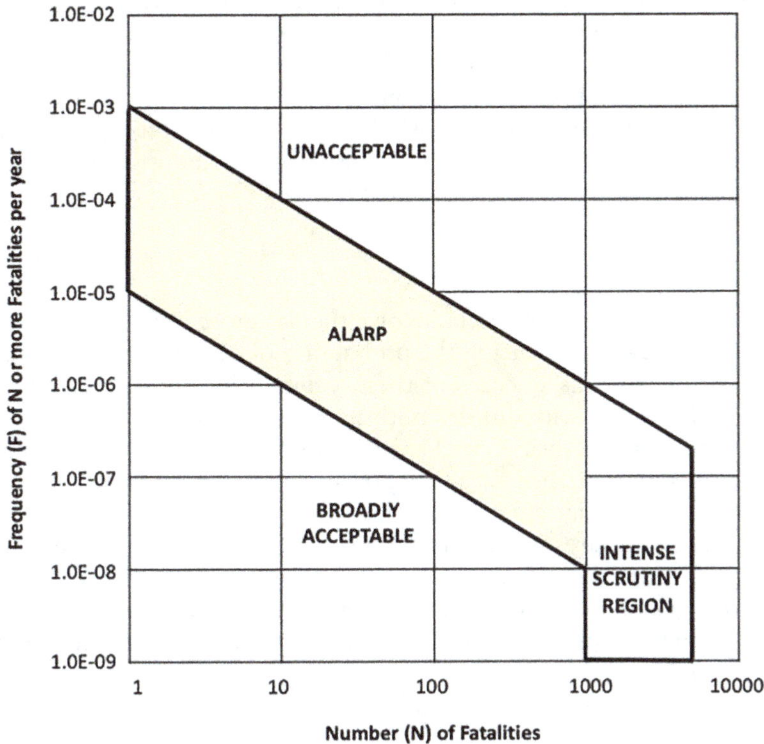

Figure 1.3 [F - N] diagram – relationship between frequency of fatalities per year (F) and number of fatalities (N) (UK Health and Safety Executive, 2010).

to the possible loss of life caused by the event [N]. The risk determined in *Task 3* may be "acceptable", or "unacceptable" or fall into an intermediate category known as "ALARP" – As Low as Reasonably Possible. ALARP applies to conditions where the cost of effective mitigation far exceeds the benefit that may be achieved by implementing the mitigation, in which case the risk would be considered as generally acceptable. A fourth "intense scrutiny region" is defined by very high-consequence events where the possible loss of life exceeds 1,000, but the annual frequency of these events is very low. A careful study of the annual frequency of these events and their consequences would provide guidance on the type and magnitude of mitigation measures that may be effective in reducing risk. Presumably, analysis of risk for the "intense scrutiny region" would examine a wide range of societal and fiscal conditions with the aim of reducing the risk to acceptable levels if the risk cannot be eliminated (UK Health and Safety Executive, 2010).

Having determined the risk in *Task 4*, the next task (*Task 5*) is to determine if the risk is acceptable based on the guidelines shown in Figure 1.3, for example.

For risks that are unacceptable, mitigation measures will be required to reduce the likelihood of occurrence and/or to reduce the consequence until the risk is acceptable (*Tasks* 6 and 7). If the risk is acceptable, then the project can proceed taking into account the mitigation measures that have been taken to achieve acceptable risk (*Task* 8). As the project proceeds, the risks identified during design would be monitored and mitigation measures updated as appropriate, or risks could be retired (*Task* 9).

1.4.4 Risk and climate change

Many risk management projects involve the design of remedial works that have the objectives of reducing the probability of occurrence of hazardous events and/or the consequences of these events. The design and operation of remedial works must consider both past conditions that influenced the hazard, and possible future conditions and how they may differ from the past due to climate change (Lacasse, 2021).

One of the most significant influences on the long-term performance of geotechnical structures is rainfall, both directly in terms of increased pore pressures in slopes, and indirectly such as flooding and scour. Design of remedial work should take into account, if appropriate, that rainfall in the future may be more intense and of greater duration than in the past. Another possible climate change effect is wild fires that expose fragile soils to rapid scour and erosion by rainfall.

1.5 EXAMPLES OF LOW-RISK AND HIGH-RISK SITES

To illustrate the use of the data shown in Figures 1.1 and 1.2 to carry out risk analysis and give examples of low-risk and high-risk conditions, two sites are discussed below. The low-risk, rock fall site refers to event risk, and the high-risk, unstable foundation site refers to an activity risk.

1.5.1 Low event risk – Uluru, Australia

Uluru (formally Ayers Rock) is a 348 m high, sandstone monolith in central Australia, located about 400 km west of Alice Springs. Uluru was formed between about 400 and 550 million years ago when immersed alluvial sediments were compacted into conglomerate, and then the horizontal strata were rotated vertically by the Alice Springs orogeny. Finally, the surrounding sediments were eroded to expose Uluru – the sandstone is strong and resistant to erosion and is also massive with no significant bedding or joint plane discontinuities (Figure 1.4).

The longevity of Uluru is a consequence of the strength and massive nature of the sandstone, as well as the hot, dry weather and seismic stability

Figure 1.4 Uluru rock formation in central Australia; image of rare rock fall at base of monolith.

of central Australia. However, rock falls do occur occasionally, but these are rare events as shown by the absence of significant talus deposits surrounding the monolith.

Risk management of the rock fall hazard events at Uluru involves examination of the following data:

i. **Probability of rock fall occurrence** – if it is estimated that, at present, the number of rock falls that have accumulated around Uluru is about 5,000. If these falls have collected over the past several million years, then on average, one fall occurs every 2,000 years or an annual frequency of 5E-4. Even if the number of falls that have occurred is greater by an order of magnitude, rock falls are very rare events.

ii. **Consequence of rock falls** – the only access to the base of Uluru is a 10.6 km long pedestrian pathway around the monolith which is not heavily used; no buildings such as hotels are close by. Therefore, the likelihood of a person being struck by a rock fall is remote, although if it were to happen, the consequence would probably be a fatality or serious injury because the dimensions of falls of this massive rock could be several metres as shown in Figure 1.4. Furthermore, rock falls will occur with no warning.

iii. **Risk Management** – while the probability of a rock fall occurring, and striking a person at the fall location (low spatial probability) and at the moment of the fall (low temporal probability), is very low. However, the consequence of such an event could be severe. Therefore, prudent management of the risk would be to restrict access to areas of active falls, if such areas are identified. However, because the rock fall risk is very low, pedestrian pathways around the base of Uluru are generally open, and for most people, it is worth the very low risk of being struck by a rockfall to walk around the base and observe the interesting geological formations developed over millions of years.

1.5.2 High activity risk – Kariba dam hydroelectric project, Zambezi river

Kariba Dam is a 128 m high, double curvature, concrete arch dam on the Zambezi River on the border between Zimbabwe to the south, and Zambia to the north (Figure 1.5). The lake created by the dam has a surface area of about 5,400 km^2, and is the world's largest man-made reservoir. Underground powerhouses in excavated rock caverns are located in the left and right banks. The project was constructed between 1954 and 1959 by the Italian company Impresit.

The spillway comprises five gates located just below the crest of the dam that discharge into a plunge pool immediately below the dam. In the early years of the project, before the north powerhouse was constructed in 1977, the spillway operated regularly. However, since 1977, the spillway has been used less frequently because most of the river flow is through the power turbines.

Despite the current infrequent operation of the spillway, it has been found that discharge from the spillway has scoured a 90 m deep hole in the plunge pool that has possibly weakened the dam foundation (Bollaert, Munodawafa, & Mazvidza, 2013). Another issue with operation of the dam is that chemical reaction between the cement and the concrete aggregate around the spillway openings has resulted in difficulty with operation of the gates.

After 60 years of successful operation of the hydroelectric development, a project was started in 2020 to correct the deficiencies in the gate operation

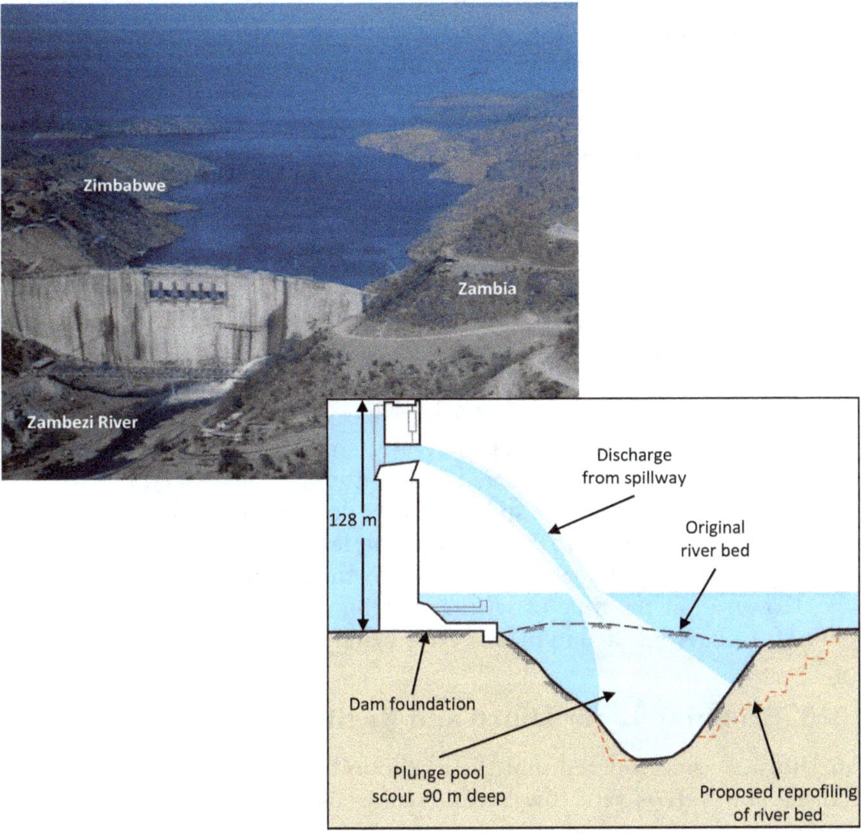

Figure 1.5 Kariba Dam on Zambezi River, central Africa (image by C. B. Wyllie); diagram of scour in plunge pool caused by spillway discharge (graphic: Institute of Risk Management of South Africa).

and the dam foundation. Work on the foundation comprises construction of a coffer dam downstream of the main dam that will allow the plunge pool to be pumped dry and the foundation inspected. It is planned that rock will be excavated to enlarge and reshape the plunge pool to reduce scour, and if necessary, the foundation rock will be reinforced with rock anchors and a concrete facing.

Risk management of Kariba Dam would require examination of the following activity issues.

- **Probability of collapse of the dam** – studies of the dam and its foundation to about 2020 show no evidence of significant deterioration of the dam itself, but the condition of the foundation will only be known when it can be inspected after the plunge pool is pumped out. Meanwhile, stability will continue to deteriorate each time the spillway operates.

- **Consequence of dam collapse** – the consequence of the dam collapse would be catastrophic, particularly if it occurred with little warning. That is, the populated Zambezi Valley between Kariba and the mouth of the river on the Indian Ocean, a distance of 1,300 km, would be severely damaged. In addition, power generation would have to resort, for many years, to coal-fired plants to replace the hydroelectric generation at Kariba.
- **Risk management of dam collapse** – with the possibility of dam instability, and the extreme consequence of such a failure, the risk is high (Darbourn, 2015). Management of the risk can only be achieved by eliminating scour of the foundation to reduce the probability of dam failure because it is not possible to reduce the consequence of this event. Regarding the planned mitigation program of reshaping the plunge pool, a further risk to be considered is that it may be necessary, due to flood conditions, to open the spillway gates during the work. Mitigation of this risk would require that the reservoir level be kept low by running the river flow through the turbines in the north and south bank powerhouses. However, at the time of writing (2023), drought conditions in central Africa have resulted in the reservoir being well below full capacity.

1.5.3 Risk matrix for Uluru and Kariba dam

For Uluru, it is considered that the likelihood of a rock fall occurring and injuring someone is very low because the rock is strong, the dry weather conditions are favourable and few people walk around the base of the site. However, the consequence of a rock fall impacting someone is high because of the large size of the falls in this massive rock. The matrix shows that the risk at the site is moderate (Figure 1.6).

For Kariba Dam, the likelihood of the dam being unstable is moderate because the dam is carefully inspected and monitored, and rehabilitation work of the foundations and spillway is in progress (2023). The consequence of instability of the dam is extreme because of the potential downstream damage. The matrix shows that the risk at the site is moderate, but higher than that for Uluru.

Figure 1.6 shows the occurrence likelihood/consequence risk matrix with the risk conditions for Uluru and Kariba Dam.

1.6 QUANTIFICATION OF UNCERTAINTY

A fundamental feature of geotechnical engineering is uncertainty in many aspects of projects; these uncertainties can include site conditions, design parameters, design methods and construction techniques. It is usual that

RISK MATRIX Uluru and Kariba		Consequence of negative outcome				
		Minimal consequence	Minor consequence	Moderate consequence	Major consequence	Extreme consequence
Likelihood of event occurrence	Very high likelihood					
	High likelihood					
	Moderate likelihood					Kariba dam instability
	Low likelihood					
	Very low likelihood				Uluru rock fall hazard	

Figure 1.6 Risk matrix for Uluru rock falls and Kariba Dam instability.

the reliability of the design will be improved if these uncertainties can be quantified such that the parameter(s) that are the most uncertain, and have the greatest influence on design, can be identified and appropriate design improvements implemented. This section discusses methods that can be used to quantify uncertainty by defining uncertainties as mathematical probability distributions, rather than incorporating contingencies into values of the design parameters.

1.6.1 Probability distributions

The first step in the quantification of uncertainty is to define the probability distribution of each design parameter. Figure 1.7 shows four typical distributions, generated by the program *@Risk* for the cohesion of a rock mass (Lumivero Corporation, 2022):

i. Triangular
ii. Normal
iii. Beta
iv. Lognormal

Cohesion was selected for this example because this parameter is often difficult to measure in the laboratory, and judgement is required to determine design values. In developing these distributions, it is assumed, based on the experience of the site, that the cohesion could have an average value of 100 kPa, but have a range of between 0 and 200 kPa. The following comments apply to these distributions:

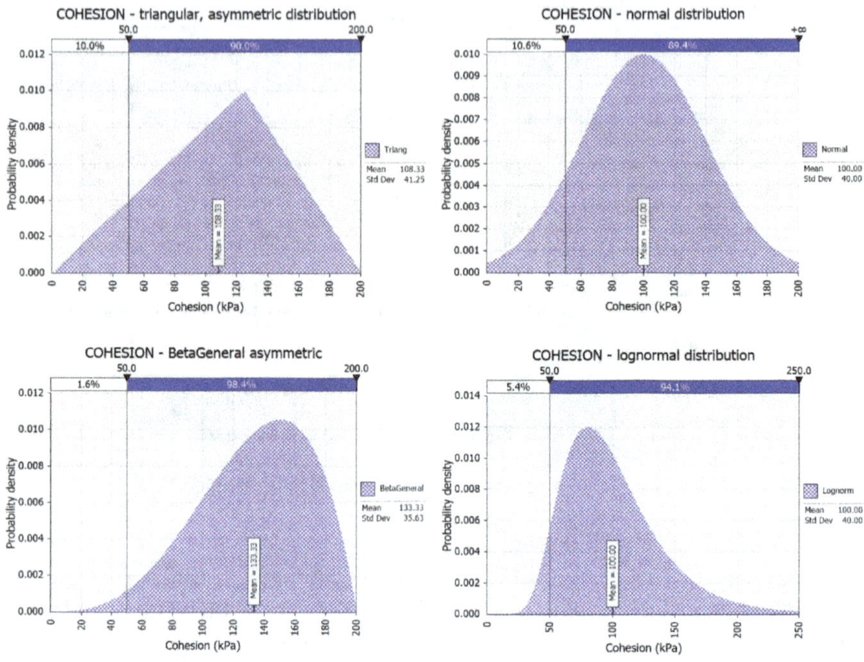

Figure 1.7 Examples of probability distributions for cohesion of a fractured rock mass—triangular, normal, beta and lognormal (plots generated by @Risk, Lumivero Corp).

a. the normal distribution, which is symmetrical and has no upper and lower limits, is a commonly used distribution and is defined by the equation:

$$p(x) = \frac{1}{\sigma\sqrt{2\pi}} exp\left[-\frac{1}{2}\left(\frac{x-\mu}{\sigma}\right)^2\right] \tag{1.2}$$

where *p(x)* is the probability of occurrence of a particular value of x, μ is the mean value of the population, and σ is the standard deviation of the population. If the population has N values and x_i is the value for each member of the population, the standard deviation is given by:

$$\sigma = \sqrt{\left(\frac{\Sigma(x_i-\mu)^2}{N}\right)} \tag{1.3}$$

Equation (1.3) can be used to calculate the standard deviation for distributions other than the normal distribution – Figure 1.7

shows calculated values of σ for Triangular, Beta and Lognormal distributions.

b. The width of the plot is related to the standard deviation, with the width of the plot increasing as the value of the standard deviation increases. That is, at higher values of σ, the uncertainty increases as shown by the wider range between the upper and lower parameter values.
c. For each of the four plots, the area under the plot is equal to 1.0, consistent with the fact that the total of all probabilities for any distribution is equal to unity. This means that the height of the distribution plot decreases with higher values of the standard deviation (σ) indicating that the data have more scatter, to maintain the area of unity.
d. For the triangular and Beta distributions, the maximum and minimum values are defined, while the lognormal distribution has a minimum value of 0 and the maximum value extend to infinity. It is considered that all design parameters have upper and lower bounds, so the triangular and Beta distributions are better able to represent actual conditions and avoid the influence of "long tails" of normal and lognormal distributions.
e. The triangular distribution, whether symmetric or asymmetric, is a simple means of defining the upper and lower bounds of a parameter, and a most likely value. This may be useful when the site information is limited or is non-existent and estimates based on experience must be used for design.
f. The Beta distribution, whether symmetric or asymmetric, is a versatile distribution that can model many design parameters.
g. The Lognormal distribution can closely approximate a normal distribution, but with the important difference that it cannot have a negative value. The lower limit is zero while the upper limit is infinite, and the plot can be skewed to suit asymmetrical data. The Lognormal distribution is a useful means to model the distribution of particle sizes in soil and rock. For example, most rock falls that reach the base of a slope will have small dimensions because blocks are fractured during the fall as they impact the slope. However, the Lognormal distribution accounts for the occasional much larger rock that must be considered in the design of protection structures.
h. For each distribution, the standard deviation is shown, with values ranging from 35–41 kPa.
i. Each plot in Figure 1.7 shows the probability that the actual cohesion will have a value of less than 50 kPa. For example, for the triangular distribution, the probability that the cohesion is less than 50 kPa is 10%. This information would be useful, for example, if it were determined that a cohesion of 50 kPa was a critical value below which the factor of safety would decrease significantly. The plots show that the normal distribution has the highest probability (10.6%) that cohesion

will be less than 50 kPa because the graph extends to infinity for cohesions less than 50 kPa, and is not constrained at the upper limit. The lowest probability of the cohesion being less than 50 kPa is for the Beta distribution that is skewed to the right indicating that high cohesion values are possible. It is considered that distributions that constrain the upper and lower limits of parameters to realistic values are preferred in analysis to distributions that extend to infinity.

1.6.2 Defining uncertainty

Development of probability distributions such as those in Figure 1.7 will require different approaches, depending on the information available, as discussed below. These approaches to developing probability distributions are discussed further in Section 2.3

a. **Data analysis** – if information exists, for example, on material strengths from laboratory tests, or of groundwater pressures from piezometer measurements, then these can be analyzed and the data fitted to the most appropriate probability distribution. If necessary to account for extreme values that may only have a low probability of occurrence, the fitted distributions could be adjusted by, for example, increasing the probability of extreme low strength values and high water pressures.
b. **Experience and judgement** – it is rare that sufficient data of a design parameter is available that can be fitted to a probability distribution with a high correlation coefficient. Where available design data is limited, the experience and judgement of the project team can be used to develop suitable distributions of the data.
c. **Expert opinion** – for large or high-risk projects where it is desirable for design to use reliable probability distributions but little site data is available, it may be necessary to convene a panel of experts to prepare information on site conditions. One approach is to use a Delphi panel where the opinions of each member are circulated anonymously to the other panel members until consensus is reached (see Section 3.4.2).

1.6.3 Black swan events

Despite the best efforts of the project team to develop realistic probability distributions, it is rare that they will consider very severe hazards or consequences that are difficult to predict and may only occur very rarely and be beyond the realm of normal expectations. Such events are sometimes called "black swans" that relates to the breed of swans that inhabit the Swan River in Perth, Western Australia; these swans are coloured black in contrast to most other swans in the world that are white.

Examples of black swan events are the September 11, 2001 attack on the World Trade Centre in New York (Silver, 2012), and the Ka Loko Dam in Hawaii that failed after 120 years of safe operation because the spillway was filled in, resulting in seven deaths (Brummund, 2011). In determining the risk of such black swan events, the combination of severe consequence but very low probability of occurrence may result in the risk being classified as moderate (see Figure 1.2).

For black swan events, acceptance of the risk will require that the risk be compared to other similar events and generally accepted societal risk. As discussed in more detail in Chapter 5, the range of risk acceptance comprises: "Unacceptable"; "As low as reasonably possible (ALARP)"; "Generally acceptable". If the black swan risk is in the ALARP range, it may be necessary to decide whether it is worthwhile to implement a high cost remedial program to reduce the risk by a minor amount if an event has a very low probability of occurrence. Regarding the Ka Loko Dam, it would be difficult and expensive to construct a spillway that could not be filled in, and the designers would reasonably expect that this would not be required because the probability of this happening would be negligible.

1.7 SOURCES OF UNCERTAINTY

In the preliminary stages of a project, it may be valuable to make an assessment of which design parameters are most critical to design, and which may have the greatest uncertainty. This information could then be used to plan the investigation program, concentrating on critical values with the greatest potential uncertainty.

The following is a discussion on common sources of geotechnical uncertainty.

1.7.1 Site condition uncertainty

Usual methods of investigating sub-surface site conditions are surface mapping, excavating test pits, drilling investigation holes and carrying out geophysical studies. The limitations of these methods are that mapping gives little information of soil and rock conditions at depth, test pit depths are restricted to a few metres, drill holes sample a very small fraction of the site, and geophysics is only an interpretation of sub-surface conditions. The development of a geological model of the site will require interpolation and extrapolation of the discrete information at each test pit and borehole, with much opportunity for uncertainty.

A study by the World Bank examined the reasons for cost overruns on power generation projects and emphasized the value of detailed site investigation programs (Hoek & Palmeiri, 1998). In addition, a study of costs for 84 tunnelling projects in the United States examined the relationship between cost increases for differing site conditions (DCS) and the extent of exploration drilling prior to contract award (U. S. National Committee on Tunnel Technology, 1984). As shown in Figure 1.8, the study found that if the ratio of the drill hole length to the tunnel length (L_{dh}/L_t) was less than 1.0, for some projects the cost overruns could be as high as 80% of the engineer's estimate although for many projects the cost increase was less than 10%. However, little data is available for L_{dh}/L_t >1.0 so it cannot be concluded that more drilling would significantly reduce cost overruns. Furthermore, the cost overrun data is scattered over almost one order of magnitude, indicative of the uncertainty that exists in the cost of tunnel projects.

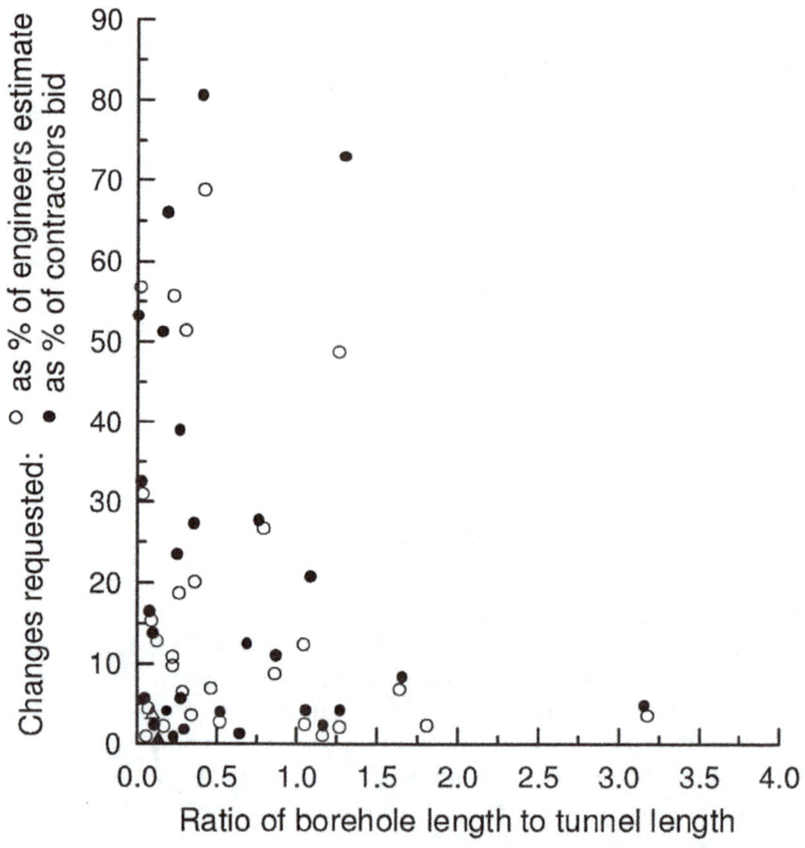

Figure 1.8 Tunnel cost overruns related to length of exploration drill holes compared to length of tunnel.

1.7.2 Material strength uncertainty

Strength properties of soil and rock that are often required for design include compressive strength, deformation modulus, cohesion and friction angle. Laboratory testing of these parameters requires the retrieval of undisturbed samples that are representative of the site, and that the laboratory-scale strength values can be scaled up to the full dimensions of the project. With respect to the strength of fractured rock masses, it is rarely possible to carry out laboratory tests of these materials and it is necessary to use indirect methods to determine the strength such as the Hoek–Brown rock mass strength criterion (Hoek & Brown, 1997).

A further source of uncertainty is that strengths can change with time. For example, cementation by deposition of calcite can increase strength, and weathering of shale can reduce its strength.

1.7.3 Groundwater pressure uncertainty

Sources of information on ground water at a site include the climate such as seasonal rainfall and snow melt, observation of seepage and measurement of piezometers at discrete locations, all of which have limited reliability. Furthermore, ground water pressures can change with time in response to the seasons, precipitation levels, installation of drainage measures and climate change.

If the design incorporates water pressure, then a value for the water pressure distribution will need to be selected that is not too conservative but still reflects the likely uncertainty.

1.7.4 Model uncertainty

Once the design parameters have been finalized, it is necessary to select a design model. For example, stability analysis of a soil or closely fractured rock mass will involve a circular failure surface where it is necessary to select the radius of the sliding surface, and the position and depth of the tension crack, if any. The values of these parameters all influence the calculated factor of safety, and while most analysis programs search for critical circles, the results may not match reality, particularly if uncertainty in design parameters is significant.

1.7.5 Seismic ground motion uncertainty

In seismic regions, designs will incorporate the effect of ground motions on slope stability and structure performance. Commonly used seismic design methods are pseudo-static analysis in which a static horizontal acceleration is applied to the slope, or Newmark displacement analysis in which slope movement in response to cyclic ground motions is calculated. Both these analysis methods are approximations of actual behaviour in the event of an earthquake – that is, applying a static horizontal acceleration is not an accurate simulation of very brief, dynamic, three-dimensional seismic acceleration, and calculated slope displacement will not be definitive information on whether a slope will fail.

1.7.6 Construction method uncertainty

Despite the preparation of thorough designs and comprehensive specifications, problems can still arise during construction that requires design modifications that may result in cost increases and schedule delays, or possibly failures and injuries. Examples of construction issues that can occur are flooding during the construction of a bridge or dam, intersection of major fault in a tunnel that results in collapse, or blasting damage to a structure or to rock that significantly reduces its strength.

Methods to mitigate construction risks would be to use risk identification studies during design to identify potential construction hazards and address these issues in all stages of the project. For example, crews should have the necessary skills, experience and equipment, and quality management plans should be implemented.

1.8 CASE STUDIES ILLUSTRATING RISK MANAGEMENT

To illustrate the application of risk management to geotechnical engineering, four diverse projects have been selected for each of which the eight tasks identified in Figure 1.1 will be applied in subsequent chapters. These eight tasks are:

i. Hazard/consequence identification
ii. Hazard/consequence analysis
iii. Risk calculation
iv. Risk evaluation
v. Risk acceptance
vi. Risk mitigation
vii. Project execution
viii. Risk monitoring

The four projects are grouped according to the following two classes of risk:

a. **Event risk** – projects where an annual probability of failure exists and it is required to estimate the likelihood of failure occurring and the possible consequences of such an event. The two case studies are first, a debris flow channel above vulnerable infrastructure, and second, rock falls on a transportation system.

b. **Activity risk** – projects where a single activity is carried out and risk exists that the project, or one component of the project, is unsuccessful resulting in cost overruns and schedule delays. Two case studies are first a tunnel intersecting a fault possibly resulting in a collapse of the tunnel, and second, installation of high-capacity rock anchors to stabilize a concrete gravity dam.

Descriptions of these four projects are provided below.

1.8.1 Debris flow case study – event risk

Debris flows are highly fluid mixtures of water, solid particles and organic matter. The mixture has a consistency of wet concrete and comprises about 40% to 50% water, and solid particles ranging from clay and silt sizes up to boulders several metres in diameter. Where such flows originate in streams with gradients steeper than about 20°–30°, they can move at velocities of 3–5 m/s, with pulses as great at 30 m/s (Skermer, 1984).

Debris flows can be very destructive, and their frequency and size are unpredictable. For this reason, risk analysis and risk management can be carried out to determine the optimum method of protecting infrastructure such as housing and transportation systems from these events. Possible remedial actions are to either remove vulnerable structures from the potential path of the flow or construct a containment structure in the creek channel (see sketch) (Jakob & Holm, 2012).

This case study of risk management for a debris flow examines the eight risk management tasks listed in Figure 1.1, as follows:

i. **Hazard/consequence identification** – in steep terrain in high rainfall areas, potential debris flow channels are usually readily identified, as well as vulnerable infrastructure in the run-out areas.
ii. **Hazard/consequence analysis** – the most challenging risk management task is to determine the likely frequency-magnitude relationship for future events; this usually includes detailed examination of the fan to examine the characteristics of past events. Risk analysis also includes estimation of the consequences of the possible range of future events.
iii. **Risk calculation** – risk for the site is calculated as the product of the likelihood of occurrence, usually expressed as an annual frequency, and the consequence expressed as lives lost (probability of death of an individual, PDI) and/or monetary cost.
iv. **Risk evaluation** – the calculated risk is compared with generally accepted societal risk standards such as [F – N] diagram (see Figure 1.3).
v. **Risk acceptance** – evaluation of the calculated risk will show if it is "Acceptable", or "Unacceptable", or lies in an intermediate range of "As low as reasonably possible (ALARP)".
vi. **Risk mitigation** – if the risk is not acceptable, then mitigation measures to either reduce the risk (e.g., construct a barrier in the creek)

or to reduce the consequence (e.g., restrict development in the run-out area) would be implemented.

vii. **Project execution** – once the risk has been reduced to an acceptable level, the project can proceed.

viii. **Risk monitoring** – over the life of the project, the acceptable risk would be maintained by ensuring that either no development occurred in the run-out zone, or that the protection structure is maintained.

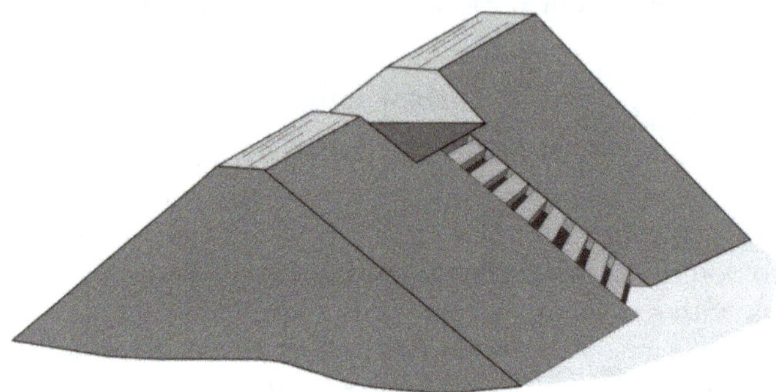

A particular risk for the construction of debris flow protection structures is that an event occurs during construction that is a hazard to personnel and equipment. This hazard can usually be mitigated by scheduling construction during periods of dry weather and interrupting the work if a sudden rainstorm occurs. An example of this situation is in Japan where a series of concrete gravity containment structures are continuously constructed on creeks that flow down the Tateyama caldera above the city of Toyama in Chubu Region. In this situation, the creeks contain a continuous bedload of sand and gravel, with occasional higher volume events that also contain boulders. Because the flow is continuous, the barriers fill as they are constructed and by the time each is completed it is at capacity and it is necessary to start another dam upstream. This situation requires particular attention to local weather conditions to avoid sudden events that do not allow time for evacuation of the site (Nomura, T., personal communication 2010).

1.8.2 Rock fall case study – event risk

Transportation systems in mountainous terrain are often subject to rock fall and landslide hazards that can result in consequences such as interruptions to traffic, damage to vehicles and injury to persons. Under these circumstances, a risk management program is often justified to optimize mitigation programs that reduce the rock fall hazard.

The following is a summary of tasks that would be carried out to manage rock fall risks.

i. **Hazard/consequence identification** – transportation systems in mountainous terrain often make an inventory of geological features such as rock and soil cuts as well as tunnels and bridge foundations. Other components of the inventory may be records of rock falls and stabilization work.

ii. **Hazard/consequence analysis** – rock fall records can provide vital information on locations and sizes of rock falls from which annual frequency-magnitude relationships can be developed, together with the consequences of these events.
iii. **Risk calculation** – where the transportation system has many potentially unstable sites, it will be necessary to rank the sites according to the level of risk and then calculate the risk for the high-risk sites. This will allow remedial work to be focused on where it will be most effective in improving operational safety.
iv. **Risk evaluation** – the calculated risk in terms of annual probability of failure and number of lives lost can be compared to the level of accepted risk on [F -N] diagram. For private railroads, a corporate risk level in terms of lost revenue due to traffic interruptions may be used to evaluate acceptable risk levels.
v. **Risk acceptance** – comparison of calculated risk with accepted risk levels will determine if mitigation measures are required.
vi. **Risk mitigation** – where mitigation is required, this will be limited to stabilization such as rock or soil reinforcement or installation of protection structures, because it is not possible to reduce consequences by redirecting or controlling traffic.
vii. **Project execution** – if stabilization is carried out, this will take time to plan the work so that it will minimize delays to traffic.
viii. **Risk monitoring** – because soil and rock weather, and lose strength over time, regular inspections will be required to ensure that the remedial work has been effective and to determine the risk of other sites where remedial work has not been carried out.

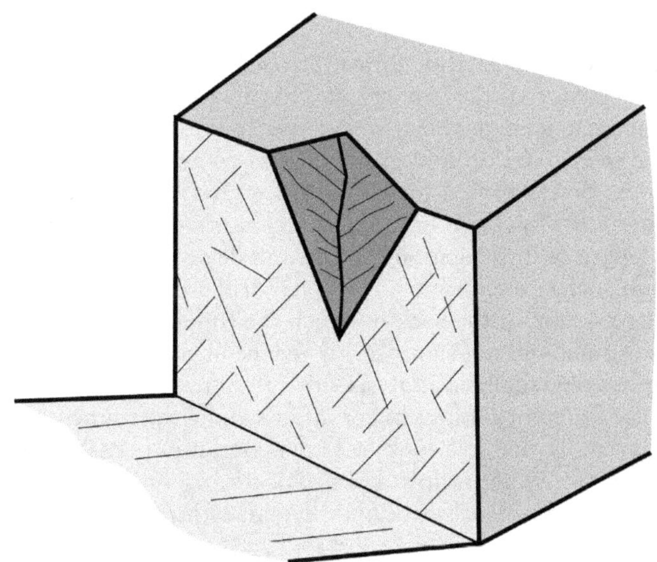

1.8.3 Tunnel stability case study – activity risk

Tunnelling is often a high-risk activity because of possible uncertainty in the geological conditions, and the need to install appropriate support to ensure safe working conditions over the full length of the tunnel. The risk will increase with increasing depth and higher *in situ* stress, and geological complexity with the possibility of the tunnel intersecting faults and contacts between different rock types. In addition, groundwater inflows in zones of weak rock can cause sudden collapse. The possible consequences of these conditions are cost overruns and schedule delays resulting from the need to install extra support such as steel sets and/or to grout water bearing features.

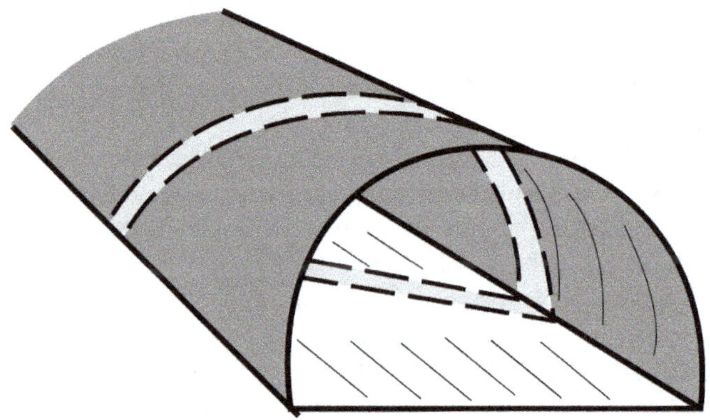

The objectives of risk management for tunnelling projects are first, to obtain as much geological information as possible (see Figure 1.8), and second, to anticipate the types and quantities of support that will be required. The eight typical risk management tasks for tunnels are discussed below.

i. **Hazard/consequence identification** – for shallow tunnels with good surface exposure of the geology, the likelihood of the tunnel intersecting unforeseen geological features is low. However, with increasing depth, more interpretation and extrapolation of available data are required, with a corresponding increasing uncertainty in the geological model. Figure 1.8 shows the negative correlation between cost overruns and the length of investigation drilling and demonstrates the possible value of sub-surface exploration. However, drill holes will only sample a miniscule fraction of the rock in which the tunnel will be driven so extensive drilling will not significantly reduce uncertainty in the geology.
ii. **Hazard/consequence analysis** – design of rock tunnels involves assessing the different geological conditions along the tunnel that may range from strong, massive rock to faults containing crushed rock that are conduits for water inflow. For each category of rock condition, a support method is developed, and then an estimate is made of the length

of tunnel that will be applicable for each support category. While this is a simple strategy for tunnel design, the possible risks are that an assumed geological condition will be incorrect, that planned support methods will be inappropriate for the conditions, and that the estimated length of tunnel for each support category will be inaccurate. For each of these risks, a potential consequence exists in terms of cost increases and schedule delays.

iii. **Risk calculation** – if numeric scores are assigned to the likelihood of tunnel instability, and to the consequence of this instability, then the product of these two scores is the tunnelling risk.

iv. **Risk evaluation** – the acceptability of the calculated risk will depend on the corporate risk tolerance for the project. For this type of one-time project, the acceptable societal risk criteria as shown in Figures 1.2 and 1.3 are not applicable.

v. **Risk acceptance** – if the calculated risk is more than the defined corporate risk threshold, then the project cannot proceed until mitigation measures have been implemented to reduce the risk.

vi. **Risk mitigation** – measures to mitigate the risk could be to carry out additional exploration in specific areas where uncertainty in geological conditions is high, or to upgrade a support type. In addition, probe holes could be drilled ahead of the face as the tunnel progresses to provide a warning of unanticipated conditions.

vii. **Project execution** – once the risk criteria have been met, the project can proceed.

viii. **Risk monitoring** – as the tunnel progresses, conditions in the high-risk areas can be monitored to determine if more or less mitigation measures are required.

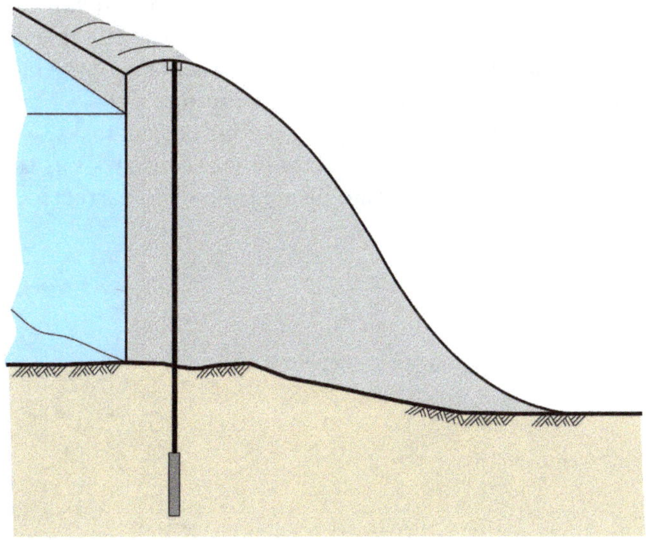

1.8.4 Dam anchoring case study – activity risk

Installation of tensioned anchors, to retain structures and support slopes for example, is a well-established and widely used operation. However, not all anchoring projects are successful, and these risks should be examined, and appropriate mitigation implemented if the risk level is unacceptable.

The eight risk management tasks for a project to install tensioned rock anchors in a concrete gravity dam are discussed below.

i. **Hazard/consequence identification** – potential issues with anchor installation include drill hole misalignment and collapse, leakage of cement grout out of the bond zone and failure of the rock-grout bond.
ii. **Hazard/consequence analysis** – the issues of drill hole stability and grout leakage are related to site geology that can be addressed from a study of the site geology, while the issues of hole alignment and anchor bonding can be addressed by using appropriate equipment and experienced work crews. The possible consequences of these negative outcomes are cost increases and schedule delays, and possible danger to personnel if the anchors are being installed for emergency stabilization.
iii. **Risk calculation** – risk can be calculated from the likelihood of negative outcomes, and the consequence of these outcomes.
iv. **Risk evaluation** – calculated risk value can be compared with corporate risk standards. For this type of one-time project, the acceptable societal risk criteria (see Figures 1.2 and 1.3) are not applicable.
v. **Risk acceptance** – if the risk is deemed unacceptable, then appropriate mitigation measures can be implemented.
vi. **Risk mitigation** – possible mitigation would be to carry out further geological investigations to check drilling and anchoring conditions and to employ experienced work crews and use suitable equipment.
vii. **Project execution** – typical tasks for an anchoring project are drilling, anchor installation, grouting and tensioning.
viii. **Risk monitoring** – monitoring of the four project tasks, with emphasis on the higher risk tasks identified in the risk analysis as the work progresses will allow problems to be identified and corrected, as required.

Chapter 2

Capacity and demand - factor of safety, probability of failure

2.1 CAPACITY AND DEMAND OF GEOTECHNICAL STRUCTURES

In general, geotechnical structures are defined by their Capacity, *C* (strength or resistance) and Demand, *D* (external loads, self-weight) that usually have differing degrees of uncertainty in their values. This chapter discusses how uncertainty in the values of *C* and *D* can be accounted for in Factor of Safety (*FS*) and Probability of Failure (*PF*) analysis. The basis of risk management is to ensure that the margin between *C* and *D* is adequate for the project circumstances. That is, the difference between *C* and *D* should be large enough that failure is very unlikely, but not too large, and expensive, that it is inconsistent with the likely consequence of failure (Christian, Ladd, & Baecher, 1994).

For geotechnical structures, Capacity and Demand are defined as follows (Nadim, Einstein, & Roberds, 2005):

- **Capacity (strength, resisting forces)** – capacity is the ability of the structure to resist deformation or instability. The primary source of resistance is usually the strength of the rock or soil in terms of the shear strength (cohesion and friction angle), compressive strength and modulus of deformation. In addition, resistance can be provided by reinforcement such as tensioned anchors in rock, or mechanically stabilized earth (MSE) in soil where the soil is reinforced with sheets of plastic grid such as Geogrid installed between layers of soil.
- **Demand (loads, displacing forces)** – loads that may be applied to a geotechnical structure. The loads may include the self-weight, groundwater pressures within the structure, and applied external loads such as bridge foundations. For a slope, the weight will depend on the slope height and face angle, the unit weight of the material, and the location of the sliding plane. The effect of water within the slope is to generate thrust forces in tension cracks and uplift forces

DOI: 10.1201/9781003271864-2

on sliding planes. Where the demand loads decrease the normal stress on the sliding plane, a corresponding decrease in the shear resistance occurs.

For a foundation, the usual design parameters are the applied bearing pressure (demand) in comparison with the allowable bearing pressure (capacity) for the foundation material. Where the foundation load is non-vertical, these demand loads will generate normal and shear components of the load in the foundation. Depending on the foundation geometry, the design may be concerned with stability or settlement of the structure.

For a structure to perform satisfactorily, its capacity/resistance must exceed the demand/load. However, as discussed in Section 1.6 above, the value of most design parameters is uncertain, and these uncertainties should be incorporated into the design. In order to meet this requirement, a number of strategies have been developed to define uncertainty in Capacity and Demand values – these strategies range from the use of judgement and experience, to quantitative methods that define each design parameter as a probability distribution.

Design strategies can be categorized as either **deterministic** in which the Factor of Safety (*FS*) is calculated using single values of each design parameter (see Section 2.2 below), or **probabilistic** in which each design parameter is assigned a probability distribution and the Probability of Failure (*PF*) is calculated (see Section 2.3 below).

2.1.1 Understanding "performance" and "failure"

All engineering designs incorporate uncertainty by using factor of safety or probability of failure criteria, depending on how uncertainty is incorporated in the calculations. In principle, the factor of safety is increased by the designer, and the corresponding probability of failure decreased, with increasing levels of uncertainty in design parameters and consequence of failure. A component of the design process is to understand the meaning of "performance" and "failure" as it applies to each specific project. For example, a highway bridge may be categorized as "unsatisfactory performance", but "safe", where settlement of a bridge abutment occurs where maintenance is required, but the settlement does not interrupt traffic and no hazard to persons exists (U. S. Army Corps of Engineers, 1997). For highway bridges, settlement may be intolerable if the horizontal movement exceeds 2 mm and the vertical settlement exceeds 100 mm (Bozuzuk, 1978; Wyllie, 1999).

In contrast to unsatisfactory performance, a design would be considered as "failed" if, for example, foundation settlement or movement was so severe that traffic could not access the bridge, or a slope collapsed resulting in damage to down-slope facilities.

The issues of failure or performance of a structure relate to the consequence of the structure's operation that should be a component of the design procedure and risk management.

2.1.2 Factor of safety guidelines

The most common criterion for designing geotechnical structures is the Factor of Safety, which, for a slope or foundation, is the ratio of the *Resisting* forces such as the material shear strength, to the *Driving* forces such as the self-weight of the slope and external loads of a bridge or building. These two categories of forces are more generally termed *Capacity* and *Demand*, respectively. The Factor of Safety incorporates all uncertainties in the design parameters into a single number, and does not explicitly quantify the uncertainties.

Values of Factor of Safety that are used in design are not prescribed in regulations, but are selected by the designer based on experience and judgement, taking into account uncertainty in the values of the design parameters, the importance of the structure and the consequences of unsatisfactory performance. Based on these criteria, Table 2.1 lists typical Factors of Safety used in geotechnical design (Terzaghi & Peck, 1967).

The factor of safety values listed in Table 2.1 were first proposed by Terzaghi and Peck many years ago, but it is considered that they have stood the test of time well based on extensive empirical experience, and are still accepted today. A good indication of the reliability of these values is that most geotechnical projects that have been designed and built using these factors of safety values have performed satisfactorily for many years. It should be noted that the values listed in Table 2.1 are based on ultimate limit states, referring to conditions where the structure fulfills the function for which it was designed - bearing capacity was not exceeded, for example. Limit states design is discussed further in Section 2.6 below.

Table 2.1 Typical Factors of Safety in Geotechnical Design

Failure type	*Item*	*Factor of Safety*
Shearing	Earthworks	1.3–1.5
	Retaining walls	1.5–2.0
	Foundations	2.0–3.0
Seepage	Uplift, heave	1.5–2.0
	Gradient, piping	3.0–5.0
Ultimate pile tests	Load tests	1.5–2.0
	Dynamic formulae	3.0

2.2 DETERMINISTIC ANALYSIS

2.2.1 Factor of safety

The most common, and simplest, method of quantifying the adequacy of a design is to determine the Factor of Safety that is defined by the ratio:

$$Factor\ of\ safety,\ FS = \frac{Capacity\ (resisting\ forces)}{Demand\ (displacing\ forces)} \tag{2.1}$$

The stability analysis method in which the Capacity and Demand ratio is calculated is termed *limit equilibrium analysis* (LEA), or *working stress design* (WSD). In this analysis, all the uncertainties in the parameter values, loads and the stability model are contained within the factor of safety. Selection of an appropriate value for each design parameter to use in the calculation of factor of safety can be based on a cautious, or conservative, estimate of the mean (Fenton & Griffiths, 2008). That is, higher, or more conservative values, would be used for shear strength parameters if they were considered to be critical to design, and little site information on their values were available.

The following is an illustration of the application of equation (2.1) to a simple rock slope stability model. Figure 2.1 shows a slope in strong sandstone containing shale interbeds with planar, persistent bedding dipping at

Figure 2.1 Image of rock slope in sandstone containing persistent bedding dipping at 30°–35°, with blocks sliding on bedding.

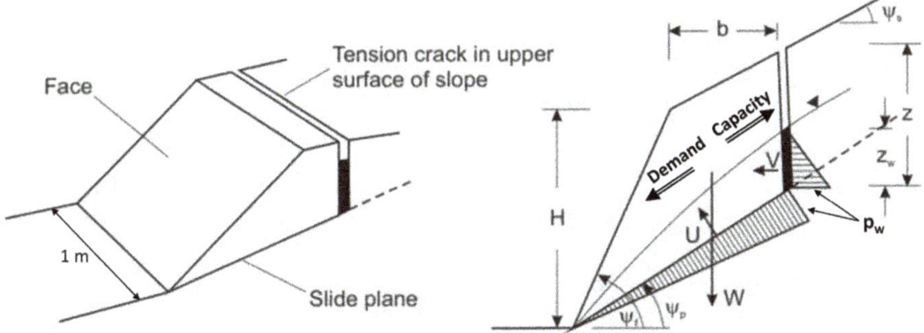

Figure 2.2 Simple sliding block model for stability calculations using Limit Equilibrium Analysis (LEA).

about 35° where blocks of rock have slid on the bedding. The stability of these blocks can be studied using *LEA* for both deterministic and probabilistic analysis using the model shown in Figure 2.2.

Figure 2.2 shows a rock slope with a height of $H = 12$ m and a face angle $\psi_f = 60°$. A tension crack with a depth of z = 6.6 m occurs at a distance of $b = 4$ m behind the crest in the upper slope that is at an angle $\psi_s = 30°$. A continuous joint dips out of the slope face at an angle of $\psi_p = 35°$. These four surfaces form a block of rock that can slide on the 35° joint plane, and is potentially unstable; the weight of the sliding block is given by the term *W*. The shear strength of the sliding plane is defined by cohesion, *c* and friction angle ϕ with values as follows :

$$c = 25\ kPa\ and\ \phi = 37°$$

If the slope is dry, and no water pressure acts in the tension crack or on the sliding plane, then the Capacity and Demand are defined as follows:

Capacity (resistance force), C provided by shear strength of the sliding plane:

$$Capacity,\ C = (c \cdot A + N \cdot tan\phi) \tag{2.2}$$

where *A* is the area of the sliding plane and *N* is the normal force acting on the sliding plane; *N* is the normal component of the block weight, *W* given by:

$$N = (W \cdot cos\psi_p) \tag{2.3}$$

Demand (displacing force), *D* is the shear component *S*, of the block weight acting down the sliding plane,

$$Demand, D = (W \cdot sin\psi_p) \tag{2.4}$$

Using these relationships for the Capacity and Demand forces, the Factor of Safety is defined as:

$$FS = \frac{(c \cdot A + W \cdot \cos\psi_p \cdot \tan\phi)}{W \cdot \sin\psi_p} \tag{2.5}$$

If the area A of the sliding plane is 13.3 m^2, and weight of the sliding block is 1,360 kN/m, then the factor of safety, FS is calculated as follows:

$$FS = \frac{(25 \cdot 13.3 + (1360 \cdot \cos 35 \cdot \tan 37))}{(1360 \cdot \sin 35)} = \frac{1182}{781} = 1.52$$

Note that the analysis is carried out for a 1 m wide slice of the slope, and all forces, both internal (e.g., weight and water) and external (e.g., applied loads), are expressed as kN/m.

2.2.2 Sensitivity analysis

The first step is assessing the influence of uncertainty in parameter values could be to carry out a sensitivity analysis to determine how variations in the value of a parameter may influence the factor of safety. For example, groundwater pressure often has a high level of uncertainty because it can only be measured at limited discrete locations, and will fluctuate with time.

Figure 2.2 shows how water pressures can be modelled approximately in a slope based on the depth of water in the tension crack z_w, and the water pressure distribution on the sliding plane. The water pressure at the base of the tension crack, and at the top of the sliding plane, are identical and have a value of:

$$p_w = \gamma_w \cdot z_w \tag{2.6}$$

where γ_w is the unit weight of water.

Assuming that the water pressure on the sliding plane drops to atmospheric pressure at the slope face, then the water uplift force U on the sliding plane, and the horizontal water force in the vertical tension crack V, are triangular distributions given by:

$$U = \frac{1}{2} \cdot p_w \cdot \text{A and } V = \frac{1}{2} \cdot p_w \cdot z_w \tag{2.7}$$

The limit equilibrium equation for the factor of safety of the slope incorporating water forces V and U is:

$$FS = \frac{Capacity}{Demand} = \frac{c \cdot \text{A} + \left(\text{W} \cdot \cos \psi_p - \text{U} - \text{V} \cdot \sin \psi_p\right) \tan \phi}{\text{W} \cdot \sin \psi_p + \text{V} \cdot \cos \psi_p} \tag{2.8}$$

Equation (2.8) shows that water forces U and V decrease the Capacity (resistance) by reducing the normal force, and shear resistance, of the sliding plane, while force V increases the Demand (driving) force by increasing the down-slope shear force. Equations (2.7 and 2.8) show that the relationship between the factor of safety and the water pressure in the slope defined by z_w is as follows:

Dry slope: z_w/z = 0%	FS = (1182/780) = 1.52
Saturated slope: z_w/z = 100%	FS = (756/959) = 0.79
Partially saturated: z_w/z = 2.33/6.66 = 35%	FS = (1055/802) = 1.32

These results show that if the factor of safety is to be greater than 1.3, the water level in the tension crack z_w should not exceed about 2.3 m.

2.2.3 Margin of Safety

The slope stability model shown in Figure 2.2 was used to calculate the factor of safety for three different water pressures that involved calculation of Capacity (resisting) and Demand (displacing) values for each case. The difference between Capacity and Demand forces is termed the Margin of Safety, M:

$$M = (C - D) \tag{2.9}$$

with M being positive when the structure is stable, and negative when the structure is unstable. For the three water pressure conditions considered for the slope, the corresponding margins of safety are:

Dry slope: z_w/z = 0%	FS = 1.52; M = (1182 – 780) = 402 kN
Saturated slope: z_w/z = 100%	FS = 0.79; M = (756 – 959) = −203 kN
Partially saturated: z_w/z = 2.33/6.66 = 35%	FS = 1.32; M = (1055 – 802) = 253 kN

The design magnitude of the margin of safety would reflect the uncertainty and the risk of the design, with M being greater for a design with a high level of uncertainty compared to a project with little uncertainty where a lower margin of safety would be appropriate.

2.3 PROBABILISTIC ANALYSIS

Following the deterministic calculation of the factor of safety for the slope discussed in Section 2.2 above, probabilistic analysis of slope stability allows the uncertainties in values of the design parameters to be explicitly incorporated into the analysis. This requires that the value of each design parameter be expressed as a probability distribution, from which the Probability of Failure can be calculated. Methods of developing probability distributions are discussed below.

2.3.1 Estimating standard deviation

The most common type of probability distribution is the normal distribution, which is a symmetrical, bell-shaped curve, defined by the mean value μ, and standard deviation σ; the distribution extends to infinity in both directions. The normal distribution can be readily used to quantify parameter uncertainty and to calculate likely performance of geotechnical structures. However, it is found that comprehensive test data from which values of μ and σ can be determined is usually only collected for large and/or complex projects. For most projects, little or no site-specific data is available, and experience and judgement are required to develop design values. While the value of μ may be established with some confidence from either test data or experience on previous projects, determination of values for standard deviation may require different strategies as discussed below (Duncan & Sleep, 2015).

Standard deviation computed from data - where site-specific test data, or data from similar sites is available, the standard deviation σ, can be calculated from the relationship:

$$\sigma = \sqrt{\left(\frac{\Sigma(x_i - \mu)^2}{N}\right)} \tag{2.10}$$

where x_i is the value for each member of the population, and μ is the mean of the population and N is the population of the data.

Standard deviation from published values - published information for geotechnical properties that are applicable to the project may be available, and it is particularly useful where information is provided on the uncertainty or scatter of the data, such as the Coefficient of Variation (*COV*), for example. The Coefficient of Variation (*COV*) is defined as:

$$COV = \frac{\textit{Standard deviation}}{\textit{average value}} = \frac{\sigma}{\mu} \tag{2.11}$$

An example of scatter in geotechnical parameters is the data listed in Table 2.2 which shows Coefficient of Variation (*COV*) values for a range of geotechnical parameters that have been obtained from laboratory measurements and *in situ* tests. The table shows that COV values are low (3%–7%) for easily measured parameters such as unit weight, and much higher (130%–240%) for parameters that are difficult to measure such as the coefficient of permeability of saturated clay. The sampling and testing conditions, which influence the variability of the test results, are not available for the data shown in Table 2.2 and the values therefore provide only an approximation of *COV* values for any specific case - the likely degree of uncertainty should be considered when applying data from published sources (Duncan & Sleep, 2015).

Additional information on soil properties and their mean, standard deviation and *COV* values is provided from an extensive testing program

Table 2.2 Values of coefficient of variation (COV) for geotechnical properties and *in situ* tests (Duncan & Sleep, 2015)

Property or in situ test result	*Coeff. of variation, COV (%)*	*Source*
Unit weight (γ)	3% to 7%	Harr (1984) and Kulhawy (1992)
Buoyant unit weight (γ_0)	0%–10%	Lacasse & Nadim (1997)
Effective stress friction angle (ϕ')	2%–13%	Harr (1984); Kulhawy (1992)
Undrained shear strength (U_s)	13%–40%	Kulhawy (1992); Harr (1984); Lacasse & Nadin (1997)
Undrained shear strength (U_s)	Clay-UU triaxial (10%–30%)	Phoon & Kulhawy (1999)
	Clay-UC triaxial (20%–55%)	Phoon & Kulhawy (1999)
Undrained strength ratio (U_s/σ_v')	5%–15%	Lacasse & Nadim, 1997
Compression index (C_c)	10%–37%	Kulhawy (1992); Harr (1984)
Reconsolidation pressure (P_p)	10%–35%	Harr (1984); Lacasse & Nadin (1997)
Coefficient of permeability of saturated clay (k)	68%–90%	Harr, 1984
Coefficient of permeability of partially saturated clay (k)	130%–240%	Harr (1984); Benson, Daniel, & Boutwell, 1999
Coefficient of consolidation (c_v)	33%–68%	Duncan & Sleep, 2015
Standard Penetration Test blow count (N)	15%–45%	Harr (1984); Kulhawy (1992)
Electric Cone Penetration Test (q_c)	5%–15%	Kulhawy (1992)
Mechanical Cone Penetration Test (q_c)	15%–37%	Harr (1984); Kulhawy (1992)
Dilatomer Test tip resistance (q_{DMT})	5%–15%	Kulhawy (1992)
Vane shear test undrained strength (S_v)	10%–20%	Kulhawy (1992)
Plastic limit	6%–30%	Phoon & Kulhawy (1999)
Liquid limit	6%–30%	Phoon & Kulhawy (1999)

comprising more than 100 borings carried out at an interchange for a new Mississippi River highway bridge in St. Louis, Missouri (Loehr, Finley, & Huaco, 2005). These tests showed low *COV* of 3%–5% for unit weight, and higher *COV* values of 75%–260% for cohesion intercept.

If the coefficient of variation (*COV*) and the mean value (μ) for a set of data are known, then it is possible to determine the standard deviation as follows:

$$\text{Standard deviation, } \sigma = (COV)\cdot(\mu) \tag{2.12}$$

Standard deviation from "Three-sigma Rule" - values for standard deviations can be estimated using the "three-sigma rule" that is based on the fact that for a normally distributed parameter, 99.73% of all values lie within plus and minus three standard deviations of the mean (Dai & Wang, 1992). That is, an extreme low value would be about three standard deviations below the mean, and an extreme high value would be about three standard deviations above the mean. Therefore, by estimating the highest conceivable value (HCV) and the lowest conceivable value (LVC), the standard deviation can be estimated from equation (2.13):

$$\sigma = \frac{(HCV - LCV)}{6} \tag{2.13}$$

Subsequent studies have shown that engineers tend to underestimate HCV and LCV values that do not fully encompass ($\pm 3 \cdot \sigma$) (Foye, Salgado, & Scott, 2006). That is, engineers can have difficulty anticipating very rare events that are beyond the realm of normal expectations. In order to improve the reliability of the three-sigma rule, equation (2.13) has been modified as follows:

$$\sigma = \frac{(HCV - LCV)}{4} \tag{2.14}$$

The three-sigma rule can be useful where estimates for standard deviation have to be based entirely on judgment, or on judgement together with meagre data (this issue is also discussed in Section 3.3 below concerning overconfidence in estimating probabilities).

2.3.2 Probability of failure - capacity and demand distributions

The limit equilibrium analysis discussed in Section 2.2 can be readily applied to probabilistic analysis in which uncertainty in the values of the design parameters can be quantified, rather than assumed as is the case for factor of safety analysis.

If uncertainties in the design parameters are defined by probability distributions, then the calculated values of the Capacity (resistance), C and Demand (displacement), D forces, as well as the Factor of Safety (C/D) and Margin of Safety ($C - D$), are also probability distributions from which the Probability of Failure (PF) can be calculated.

Figure 2.3 shows the typical normal distributions for Capacity and Demand for slope stability analysis. The Demand distribution, with well-defined parameters comprising mainly the slope dimensions, has a mean value of 800 kN and a standard deviation of 55 kN. This is a relatively narrow curve compared to the Capacity distribution that is a wider curve

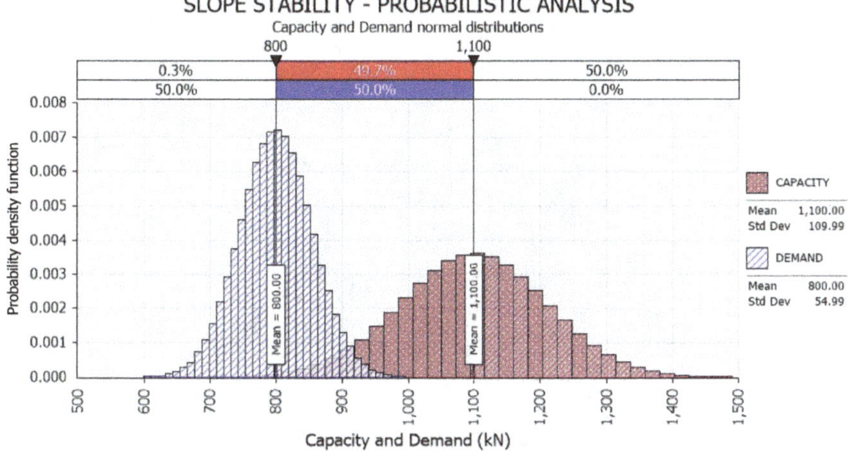

Figure 2.3 Slope stability – capacity (resistance) and demand (displacement) normal distributions for slope stability analysis (plots generated by @Risk, Lumivero Corp).

with a mean value of 1,100 kN and a standard deviation of 110 kN. The Capacity distribution has a greater standard deviation than the Demand distribution because the Capacity includes the shear strength properties of the sliding surface, and the groundwater pressure, which all have significant uncertainties.

For reference, the coefficients of variation (COV) for these distributions are:

$$COV_{demand} = \left(55/800\right) = 7\%; \; COV_{capacity} = \left(110/1100\right) = 10\%$$

That is, the distribution with the higher standard deviation also has the higher COV.

Figure 2.3 shows that the two distributions overlap for high probability values of Demand and low probability values of Capacity, such that the area of the overlap represents the probability of slope failure. The area of overlap between the two distributions can be found using the following procedure. If the Capacity and Demand normal distributions are divided, the result is a new normal distribution for the Factor of Safety (FS) where the portion of the curve where FS < 1.0 represents the Probability of Failure of the slope, at 0.7% (Figure 2.4).

That is, the Capacity and Demand distributions shown on Figure 2.3 can be used to find a probability distribution for the Factor Safety given by:

$$FS = \frac{Capacity\ (resistence)}{Demand\ (displacement)} \tag{2.15}$$

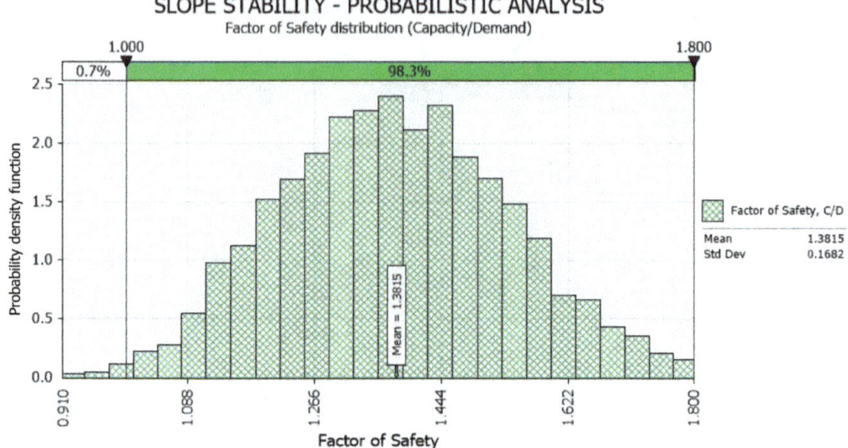

Figure 2.4 Slope stability - factor of safety distribution (factor of safety = capacity/demand) showing probability of failure (plot generated by @Risk, Lumivero Corp).

Figure 2.4 shows the Factor of Safety distribution generated by dividing the Capacity and Demand distributions where the mean factor of safety is 1.38, for which the probability of failure is about 0.7% indicating that a *FS* value of 1.38 is likely to be a stable slope because the Demand parameters are well defined. If the Demand parameters were less well defined, such as the dimensions of the sliding mass were uncertain, then the Demand curve would be wider (with a reduced maximum PDF value) with a greater area of overlap, and a greater *PF*.

The probability of failure can also be calculated from the Capacity and Demand distributions using the Margin of Safety (capacity – demand) as shown in Figure 2.5 where the Margin of Safety plot is found by subtracting the Capacity and Demand distributions. Figure 2.5 shows the mean and standard deviation values for the Margin of Safety (MS), and the Probability of Failure (PF) where the portion of the margin of safety curve is negative (MS < 0.0) – in this case PF = 0.7%.

The calculated *PF* values shown in Figures 2.4 and 2.5 may be slightly different because the curves are generated by a random number generator that produces slightly different values for each run.

It is important to note that the calculated "Probability of Failure, *PF*" shown in Figures 2.4 and 2.5 is NOT the annual probability of failure. That is, the *PF* is calculated from the same data that is used to calculate the Factor of Safety and this data contains no time information – probability of failure is just an expression that is used for the area of the negative portion Margin of Safety curve. If information on annual probabilities is required, it will be necessary to collect data on previous events. These issues are discussed in more detail in Section 2.4.

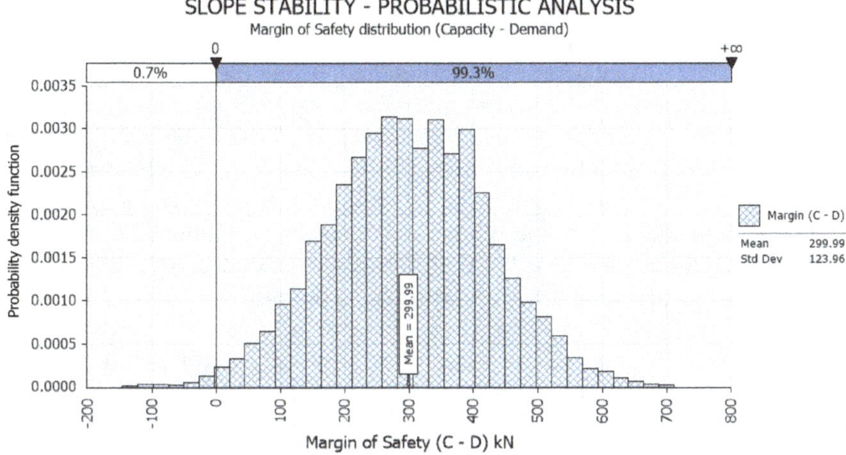

Figure 2.5 Slope stability - normal distribution of margin of safety = (capacity – demand). (Plot generated by @Risk, Lumivero Corp.)

The probabilistic stability analysis shown in Figures 2.3–2.5 has used normal distributions for the Capacity and Demand values, and the software @Risk was used to generate the normal distributions defining the *FS*, *PF* and Margin of Safety. As an alternative to the use of programs such as @Risk, it is possible to combine normal distributions deterministically as follows where μ_1 and μ_2 are the means, and σ_1 and σ_2 are the standard deviations of the two distributions. If distributions 1 and 2 are combined by subtraction, then the mean μ_r and standard deviation σ_r of the resultant distribution are respectively:

$$\mu_r = (\mu_1 - \mu_2) \text{ and } \sigma_r = \sqrt{\sigma_1^2 + \sigma_2^2} \tag{2.16}$$

For the Capacity and Demand distributions shown in Figure 2.3, μ_r = (1100 – 800) = 300 kN, and σ_r = $(110^2 + 55^2)^{0.5}$ = 123 kN. These calculated values of μ and σ closely match the values generated by @Risk probabilistic software shown in Figure 2.5.

Note that each probability distribution curve has an area of unity (1.0) because they represent all possible values for the parameter, such that a narrow distribution has a greater probability density function at the most likely value than a wider distribution.

2.3.3 Probability of failure using three-sigma rule

The procedure for calculating the Probability of Failure, discussed in Section 2.3 above, used normal distributions with assumed values for the means and standard deviations defining the Capacity and Demand.

The calculations used the program @Risk to combine the distributions for each parameter.

The following is a similar probability of failure calculation, but in this case the standard deviations are calculated using the Three-sigma Rule (equations 2.13 and 2.14) with reference to the slope model shown in Figure 2.2. The Three-sigma rule uses estimated values of the Lowest Conceivable Value (*LCV*) and the Highest Conceivable Value (*HCV*), where the *LCV* and *HCV* are determined using the parameter values for the slope model shown in Figure 2.2. For the planar slope model in Figure 2.2, the *LCV* and the *HCV* incorporate likely ranges of values for cohesion, friction angle and water pressure that are the parameters for which uncertainty is the most significant. The values for these three parameters can be difficult to define precisely because of the spatial variation across the site in the case of strength properties, and both spatial and temporal variation in the case of groundwater pressure. In contrast, the dimensions and geometry of the slope are often better defined and can be expressed as distributions with relatively small standard deviations, or as discrete values.

Referring to Section 2.2.1 and Figure 2.2 above for a simple planar stability model, equation (2.8) can be used to calculate deterministically the most likely values of the Capacity and Demand, as well as the factor of safety. The assumed most likely values for the shear strength and groundwater level are:

$c = 25$ kPa;
$\phi = 37°$
Water in the tension crack at $z_w/z = 2.33/6.66 = 0.35$

For these parameter values, the corresponding Capacity and Demand values, and the deterministic Factor of Safety (*FS*) is:

Capacity = 1,055 kN; Demand = 802 kN; $FS = (1{,}055/802) = 1.31$.

A probabilistic analysis of the slope stability can be carried out by estimating the standard deviations of the Capacity and Demand using the Thee-sigma rule (equation 2.14) discussed in Section 2.3.1 above.

The standard deviations for these Capacity and Demand normal distributions can be calculated from equation (2.14) by estimating the lowest conceivable and highest conceivable values for the three parameters with the greatest uncertainty – cohesion, friction angle and groundwater level.

Capacity - Lowest Conceivable Value (*LCV*) for capacity calculated from equation (2.8) assuming that cohesion and friction angle have very low values of:

$c = 0$
$\phi = 20°$, and

water in the tension crack has a high value $z_w/z = 6.66/6.66 = 1.00$ (saturated).

For these values, LCV for the Capacity calculated using equation (2.8), $LCV_C = 202$ kN.

Highest Conceivable Value (**HCV**), strength and water pressure values are assumed to be:

$c = 50$ kPa;
$\phi = 50°$, and
slope is dry with no water in the tension crack, $z_w/z = 0$

For these values, the HCV for Capacity using equation (2.8) is $HCV_C = 1996$ kN.

Using the Three-sigma Rule, the standard deviation of the Capacity is:

$$\text{Capacity}: \ \sigma_C = \frac{(1996-202)}{4} = 448 \ kN$$

The corresponding LCV and HCV values for Demand are $LCV_D = 780$ kN and $HCV_D = 959$ kN, from which the Demand standard deviation is $\sigma_D = 45$.

It is noted that Demand standard deviation, σ_D is smaller than the Capacity standard deviation, σ_C because the Demand is given by the term:

$$\text{Demand} = [W \cdot \sin(\psi_p) + V \cdot \cos(\psi_p)]$$

in which the shear component of the water pressure in the tension crack, $[V \cdot \cos(\psi_p)]$ is the only variable. The water pressure component of the Demand is a small portion of the total Demand force so the standard deviation is a relatively small number.

In contrast, the Capacity is given by the term:

$$\text{Capacity} = [c \cdot A + \left(W \cdot \cos \psi_p - U - V \cdot \sin \psi_p\right) \tan\phi]$$

that contains values for cohesion, friction angle and water pressure, all of which have some uncertainty.

Figure 2.6a shows the overlayed normal distributions of the Demand and Capacity forces. Although the Demand distribution is relatively narrow, it does fully overlap with the Capacity distribution indicating that the slope has a finite probability of failure. The probability of failure can be calculated by dividing the Capacity and Demand distributions to generate a new distribution for the Factor of Safety [FS = (Capacity/Demand)]. Figure 2.6b shows the Factor of Safety distribution generated by dividing the two distributions shown in Figure 2.6a. In Figure 2.6b the mean factor of safety is 1.31, which is comparable to the Factor of Safety of 1.38 shown in Figure 2.4. However, the probability of failure is 29%, compared to 0.7% in Figure 2.4, because of the comparatively large standard deviation of 449 kN assumed for the Capacity using the three-sigma rule. In Figure 2.3, the standard deviation of the Capacity is 110 kN.

(a)

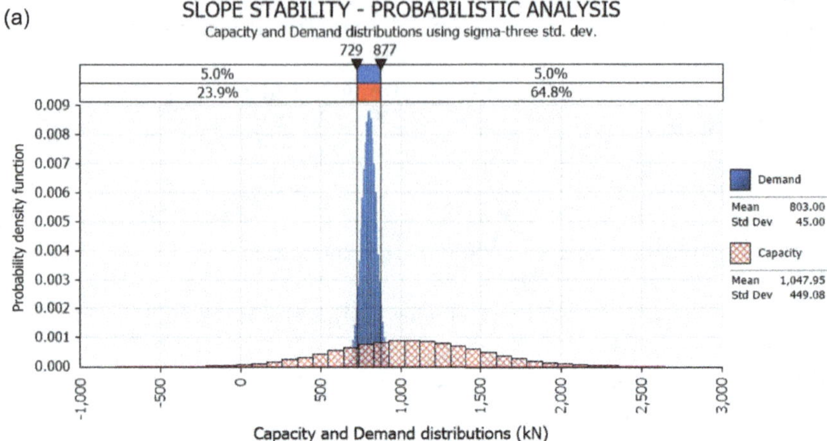

(b)

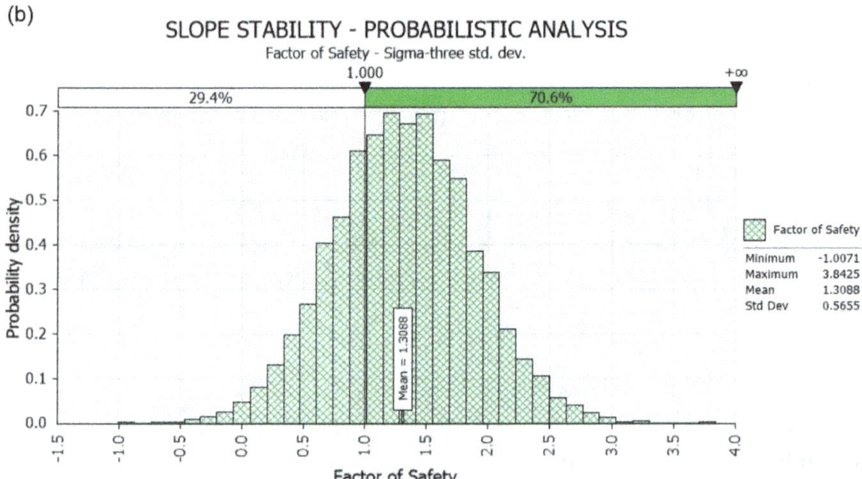

Figure 2.6 (a) Overlayed demand and capacity normal distributions using Sigma-three rule for standard deviations; (b) factor of safety distribution for Sigma-three standard deviations (plots generated by @Risk, Lumivero Corp).

2.3.4 Monte Carlo analysis

The probability of failure calculation described in Section 2.3.2 above, assumed that the values for the Capacity and Demand forces were normally distributed defined by their means and standard deviations. It is straightforward to divide these two distributions to obtain a new normal distribution with μ and σ values for the factor of safety, and subtract the two distributions to obtain a new normal distribution with σ and μ values for the probability of failure.

However, often design parameter uncertainty is defined by probability distributions such as Beta, lognormal and triangular that are a better fit of the actual data than a normal distribution. Incorporation of varied distributions into stability calculations, which cannot be carried out deterministically, may be solved using Monte Carlo analysis.

Monte Carlo analysis involves carrying out a large number of calculations (at least several thousand) for the factor of safety, for each of which a random number generator selects a value for every design parameter from their probability distributions. That is, each of the factor of safety calculations produces a unique result, which, when combined, produces a probability distribution for the factor of safety, the shape of which incorporates all the input distributions. The probability of failure is determined by counting the number of analyses where the calculations give a factor of safety less than unity. For example, if 5,000 calculations for the factor of safety are carried out and 152 of them have *FS* values less than 1, then the probability of failure is 152/5,000 = 3%.

Monte Carlo analyses can be performed using a program such as @Risk™, or similar program such as Crystal Ball, which incorporate a variety of probability distributions that can be customized to best fit the parameter values, either measured or assumed. The number of calculation iterations can be selected so that a stable result is produced.

2.4 MEANING OF TERM "PROBABILITY OF FAILURE"

2.4.1 Probability of failure calculations

Figures 2.3 and 2.4 show conditions where Demand exceeds Capacity when the probability distributions overlap, or where the distribution of the factor of safety is negative. This area of overlap, and the area of negative factor of safety, is termed the "probability of failure" (*PF*). However, this value of the *PF* is calculated from the same data as that is used to calculate the factor of safety, and this data contains no time information. That is, "probability of failure" is just an expression that is used for the area of the negative portion of the Margin of Safety curve.

2.4.2 Annual probability of failure

If information on the "annual (or temporal) probability of failure" is required, it is necessary to collect data on previous events over sufficient length of time to develop a statistically valid probability distribution. For example, many railway companies and highway authorities operating in mountainous terrain collect data on slope instability to identify high risk locations and plan effective stabilization programs. These databases of slope instability usually also contain information on the geometry, geology and topography of the sites, as well as traffic interruption data and stabilization records. Such records would form part of an overall risk management plan, not only to develop a stabilization program, but also to demonstrate that a plan is in place and being implemented, although resources are insufficient to eliminate all risk.

2.4.3 Examples of annual probability of failure data

The following are three examples of data collected to establish empirical values for annual (temporal) probabilities of failure.

a. **Talus deposits** – for glacial areas, talus deposits have accumulated since the end of the last ice age and retreat of the glaciers, an estimated time of about 10,000 years. Detailed study of a talus deposit at the base of a 320 m high, 750 m long granite cliff, determined probability distributions for both the volume of the talus deposit, and the size distribution of the talus blocks. From this information it was possible to calculate a probability distribution for the total number of falls over 10,000 years. Based on this data, the calculated average number of falls per year was about three, with a distribution range of 1–15. This study was used to assess the hazard to personnel working just beyond the outer limit of the deposit, and compare this with the [F-N] chart to ensure that the risk during the work was within the Acceptable zone (see Figure 1.3). See also Section 5.2.5 on risk acceptability.
b. **Transportation** – for a transportation system located in a mountainous area with very cold winters, frequent freeze-thaw cycles and occasional heavy rain, sample rock fall records are:

   Site 1, 2 km length – 17 slides in 36 years for an annual probability of slides of 0.24 slides/km.

   Site 2, 38 km length – 682 slides in 31 years for an average of 22 slides per year, or an annual probability of 0.6 slides/km.

   These were both high hazard locations where risk management involved extensive stabilization work, and installation of protection structures and warning systems to allow traffic to operate continuously with few delays.
c. **Project performance** – collection of information on the performance of about 75 projects over a 40 year time period has provided valuable information on the annual probability of failure related to the standard of engineering employed on the projects (Altarejos-Garcia, Silva-Tulla, Escuder-Bueno, & Morales-Torres, 2017; Silva, Lambe, & Marr, 2008).

   Two types of information were collected on these 75 projects.

A. **Characteristics of engineering program**

   As shown in Table 2.3, the characteristics of engineering programs can be defined according to five sets of project information:

   a. **Investigation**
   b. **Testing**
   c. **Design**
   d. **Construction**
   e. **Operation and monitoring**

Table 2.3 Characteristics of engineering programs

Level of engineering	*Investigation*	*Testing*	*Analysis and documentation*	*Construction*	*Operation and monitoring*
I (Best) Facilities with high consequence of failure	Evaluate design/performance of nearby structures, analyze aerial images, subsoil profile with continuous sampling, obtain undisturbed samples for laboratory testing, measure field pore pressure	Laboratory tests on undisturbed samples at field conditions, index tests to detect soft, wet, loose, high/low permeability zones	Determine *FS* using effective stress parameters based on measured data, prepare flow net for instrumented sections, prepare detailed design report	Full-time supervision by qualified engineer, construction control tests, detailed construction reports	Complete performance monitoring, comparison between predicted and measured performance, continuous maintenance
Score	0.2	0.2	0.2	0.2	0.2
II (Above average) Ordinary facilities	Evaluate performance of nearby structures, site-specific exploration program by qualified engineer	Standard laboratory tests of undisturbed samples	Determine *FS* using effective stress parameters and pore pressure	Part-time engineering supervision	Periodic inspections by qualified engineer, routine maintenance
Score	0.4	0.4	0.4	0.4	0.4
III (Average) Unimportant or temporary facilities with low consequence of failure	Evaluate performance of nearby structures, use existing subsurface data	Index tests on samples from site	Rational analyses using parameters inferred from index tests	Informal construction supervision	Annual inspection by qualified engineer, no field measurements, emergency repair only
Score	0.6	0.6	0.6	0.6	0.6
IV (Poor) Little or no engineering	No field investigation	No laboratory tests on samples collected on site	Approximate analyses using assumed parameters	No construction inspection by a qualified engineer, no construction control tests	Occasional inspections by non-qualified personnel, no field measurements
Score	0.8	0.8	0.8	0.8	0.8

Excerpted from Altarejos-Garcia, Silva-Tulla, Escuder-Bueno, & Morales-Torres (2017).

Each set of information is described according to the category of the project from Category I (Best engineering) to Category IV (Poor engineering) – see B) below. For example, investigations may range from **Category I projects:** study of nearby structures and aerial photographs, continuous core drilling, sampling for laboratory tests, piezometer installation, to **Category IV projects:** no investigation.

The scores in Table 2.3 are used to define the project category. For example, if a project is designed and constructed according to state-of-the-art practice but operation and monitoring is neglected, the total project score would be: (0.2 + 0.2 + 0.2 + 0.2 + 0.8 = 1.6) and the project could be designated as Category II.

B. Category of project

I. **Best** – facilities designed, constructed and operated with state-of-the-practice engineering. Facilities with high consequence of failure – Score = 1.

II. **Above average** – facilities designed, constructed and operated with standard engineering practice. Many ordinary facilities are in this category – Score = 2.

III. **Average** – facilities without site-specific design and with substandard construction or operation. Temporary facilities, and those with low consequence of failure, are often in this category – Score = 3.

IV. **Poor** – facilities with little or no engineering – Score = 4.

2.4.4 Documentation of annual probability of failure

The data collected for the 75 projects according to the site conditions listed in Table 2.3 is summarized in Figure 2.7 where four curves have been fitted to show the approximate relationship between factor of safety and temporal probability of failure for the four project categories listed in Table 2.3.

Annual temporal probabilities discussed in this section for talus slopes, railway/highway slopes and the 75 projects shown in Figure 2.7 demonstrate that annual probabilities of failure can range from several events per year for natural slopes, to 10^{-8} for Category I, state-of-the-practice engineering, projects. The talus deposit and railway/highway slopes can be categorized as follows according to the scores listed in Table 2.3:

a. **Talus** location is a natural slope with no "engineering" – [total score = (5 · 0.8 = 4)] - could be considered as a **Category IV** condition. The factor of safety of safety of natural slopes where slides are common is close to 1 – for these conditions, a fall may occur every few years and the annual probability of failure ≈1.
b. **Railway/highway slopes** may have been constructed many years ago with minimal engineering involvement, but since then stabilization work has often been carried out with significant design, construction

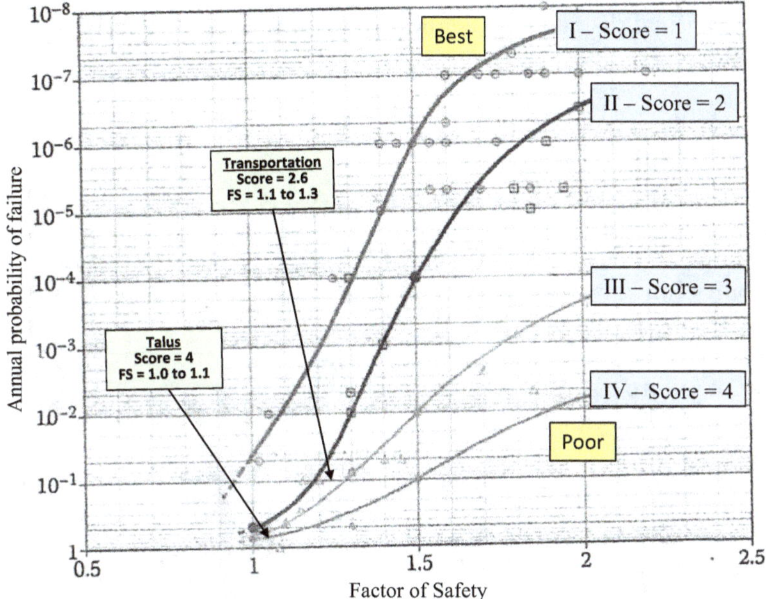

Figure 2.7 Factor of safety versus annual probability of failure for four categories of projects: I - Best to IV - Poor, listed in Table 2.3. (Altarejos-Garcia, Silva-Tulla, Escuder-Bueno, & Morales-Torres, 2017).

inspection and monitoring of performance – [total score = (0.8 + 0.6 + 0.4 + 0.4 + 0.4 ≈ 2.6)] - for which current slopes may be considered as a **Category III** condition. Records show that slides occur regularly at a rate of between about 1–0.1 per year per kilometre length.

The Factor of Safety/annual Probability of Failure relationships for these two project categories are shown on Figure 2.7.

The diversity of data shown in Figure 2.7, encompassing eight orders of magnitude in the annual probability of failure for factors of safety ranging from 1.0 to 2.0, demonstrates the difficulty in estimating *PF* design values for new projects. Another issue is the units to use for the annual probability of failure – slides/km or slides/cut slope, for example. However, the approach shown in Table 2.3 and Figure 2.7 for categorizing projects and their annual probability of failure would appear to be helpful.

2.5 RELIABILITY INDEX (β)

An additional method of incorporating uncertainty into geotechnical design is to use the Reliability Index, β that is the number of standard deviations between the most likely value of the Factor of Safety and a Factor of Safety of 1.0. The Reliability Index is also uniquely related to the Probability of Failure.

For a normal distribution defined by the mean μ and standard deviation σ, the Reliability Index, β is:

$$\beta = \frac{(\mu_{MLV} - 1)}{\sigma} \qquad (2.16)$$

where μ_{MLV} is the most likely value of the Factor of Safety. For example, the normally distributed factor of safety shown in Figure 2.4 has a mean value is 1.38 and standard deviation of 0.17. Therefore, for this distribution, the Reliability Index is: [β = (1.38 – 1)/0.17 = 2.24].

The reliability index is directly related to the probability of failure, and for a normally distributed factor of safety as shown in Figure 2.4, the probability of failure corresponding to a reliability index of $\beta = 2.0$ is $PF = 2.3\%$ see Table 8.1 and equation (8.3).

More details on the reliability index are discussed in Chapter 8 below, where methods to design structures to a specified reliability index are demonstrated.

2.6 LIMIT STATES DESIGN, LOAD AND RESISTANCE FACTOR DESIGN (LRFD)

A step towards incorporating uncertainty into design of geotechnical structures is *limit states design,* the implementation of which is termed *load and resistance factor design* (LRFD) The limit states design method was developed in structural engineering, and it is intended that the use of limit states design in geotechnical engineering will enhance consistency between structural and geotechnical designs (AASHTO, 2015). The following is background to limit states design methods.

2.6.1 Allowable stress design

The most commonly used design method in geotechnical design is allowable stress design (*ASD*). *ASD* attempts to ensure that for certain applied service loads on a structure or foundation, the stresses induced in the soil or rock mass are less than a certain set of specified allowable stresses. *ASD* combines uncertainties in loads and soil strengths (or deformations) into a single factor of safety, *FS*. That is, a set of design loads, Q_i, comprising the actual forces estimated to be applied to the structure, is balanced against a set of resistances *R* from the soil or rock bearing materials, such that:

$$\frac{R}{FS} = \Sigma Q_i \qquad (2.17)$$

or $FS = \dfrac{R}{\Sigma Q_i}$

where ΣQ_i is the total of the applied loads.

The factor of safety may be a variable that reflects uncertainty in the loads and resistances, or it may be a constant that reflects the target value for the ratio of the predicted resistances to the predicted loads. In practice, the factor of safety depends mostly on the level of control that is used in the design and construction phases of a project, and thus on resistance because uncertainties in material strengths are the greatest source of uncertainty in geotechnical design (Baecher & Christian, 2003).

2.6.2 Ultimate and Serviceability Limit States

The limit states are those conditions where the structure ceases to fulfill the function for which it was designed (Fenton & Griffiths, 2008). The two limit states are defined as:

- **Ultimate limit states** - states concerned with safety are called *ultimate limit states* that include load-carrying capacity such as bearing failure, overturning, sliding and loss of stability.
- **Serviceability limit states** - states that restrict the intended use of the structure are called *serviceability limit states* that include deflection, permanent deformation and cracking.

On this basis, geotechnical structures must satisfy both design criteria – a foundation must be designed against serviceability failure such as excessive settlement, as well as against ultimate failures such as bearing capacity failure.

At the ultimate limit state, the Factor of Safety is defined by:

$$FS = \frac{Ultimate\ capacity}{Demand} \tag{2.18}$$

In practice, the ultimate capacity has usually been found from the conservative or cautious estimates of the soil or rock properties which means that equation (2.18) is essentially the mean factor of safety. Table 2.1 in Section 2.2 above lists typical empirical factors of safety for different types of structures (Terzaghi & Peck, 1967). These Factor of Safety values refer to the ultimate limit state which is consistent with experience that very few failures have occurred historically when using the information in Table 2.1.

2.6.3 Load and Resistance Factor Design (LRFD)

As discussed in Section 2.1.2, calculation of the Factor of Safety combines all the uncertainties in the design parameters into a single value that does not quantify different levels of uncertainty in each parameter. This deficiency can be overcome to some extent by using Load and Resistance Factor

Design (LRFD) where appropriate factors are applied to each design parameter so that the Factor of Safety meets a target limit state or an acceptable Reliability Index. The use of LRFD in geotechnical design brings these designs into compliance with structural design where factored designs are well developed, partly because uncertainties in the design parameters for structural design are better defined than in geotechnical engineering (Transportation Research Board (TRB), 2009).

The principle of limit states design (and LRFD) is that Load (demand) and Resistance (capacity) parameters are multiplied by factors that are proportional to the degree of uncertainty in the parameter values. That is, a factor of greater than 1.0 is applied to the loads, with the live load factor being more than the dead load factor, assuming that the dead load is more precisely defined than the live load. Similarly, the capacity parameters such as the shear strength - cohesion and friction angle - are multiplied by factors less than 1.0.

If a design is based on the concept of resistance (capacity), R and load (demand), L, then stability will be achieved when:

$$\text{Margin, } M = (R - L) > 1.0, \text{ or } R > L \tag{2.19}$$

For LRFD design, a resistance factor φ_g is applied to the resistance, and a load factor γ is applied to the load such that the design satisfies the inequality:

$$\varphi_g \cdot R \geq \gamma \cdot L \tag{2.20}$$

Typically, the resistance factor φ_g, has a value less than 1.0 to reduce the resistance to a less likely *factored resistance* that has a small probability of occurrence, and the load factor γ is greater than 1.0 that increases the load to a *factored load* that may occur in a very small fraction of similar design situations. Regarding resistance factors, a single total resistance factor can be used to cover both cohesion and friction angle, or separate partial factors can be applied to cohesion and friction angle, with a lesser factor usually applied to cohesion because it is more difficult to measure compared to the friction angle (see Table 2.4).

A more general form of equation (2.20), which applies separate load factors γ_i to each of m types of loads is:

$$\Phi_g \cdot R \geq \eta \cdot \sum_{i=1}^{m} \gamma_i \cdot L_i \tag{2.21}$$

where L_1 may be the dead load, L_2 the live load and L_3 the environmental load such as wind. Each of these load types will have its own distribution

Table 2.4 Comparative Values for Load and Resistance Factors Used in LRFD to Calculate Footing Area (Fenton & Griffiths, 2008)

Code	*Year*	*Dead load*	*Live load*	*Friction tan ϕ'*	*Cohesion c'*	*Bearing*	*Sliding*	*Footing area (m²)*
CFEM	1992	1.25	1.5	0.8	0.5–0.65	-	-	5.2
NCHRP 343	1991	1.3	2.17	-	-	0.35–0.6	0.8–0.9	4.9
NCHRP 12–55	2004	1.25	1.75	-	-	0.45	0.8	4.7
Denmark	1965	1.0	1.5	0.8	0.57	-	-	4.5
Hansen, B	1956	1.0	1.5	0.83	0.59	-	-	4.1
CHBDC	2000	1.25	1.5	-	-	0.5	0.8	4.0
AS 5100	2004	1.2	1.5	-	-	0.35–0.65	0.35–0.65	3.9
AS 4578	2002	1.25	1.5	0.75–0.96	0.5–0.9	-	-	3.9
Eurocode 7 Model 1		1.0	1.3	0.8	0.8	-	-	3.1
Eurocode 7 Model 2		1.35–1.5	1.5	-	-	0.71	0.91	3.0
ANSI A58	1980	1.2–1.4	1.6	-	-	0.67–0.83	-	2.8

CFEM, Canadian Foundation Engineering Manual; AS, Australian Standard; NCHRP, National Cooperative Highway Research Program; ANSI, American National Standards Institute; CHBDC, Canadian Highway Bridge Design Code.

with a corresponding load factor, although load factors are typically dictated by structural codes.

The parameter η is the importance factor, which has higher values for critical structures such as hospitals and fire stations. For bridge design, the importance factors are 1.1 for lifeline bridges, 1.0 for bridges carrying medium to heavy traffic volumes and having potential impacts on alternative transportation corridors, and 0.85 for non-lifeline structures (Canadian Standards Association, 2019).

Resistance factors to use in design are not defined by code because of the variability in site conditions, but calibrations have been carried out to check that LRFD designs produce similar results to that of working stress design and limit equilibrium analysis. Table 2.4 lists Load and Resistance factors listed in geotechnical design codes from a number of different countries, with the earliest ones being in the 1950's and 1960's. Regarding the Resistance Factors, the codes list either partial factors for friction angle and cohesion, or a total Resistance Factor to be applied to the material strengths for bearing or sliding, depending on the model.

2.6.4 Examples of LRFD calculations – foundation, rock slope

The following are examples of LRFD calculations for a spread footing design, and for rock slope stability.

Foundation – Table 2.4 lists LRFD design parameters for a spread footing on soil as prescribed by several design codes. In order to compare the listed codes, the required area of a spread footing designed against bearing failure (ultimate limit state) using a dead load of 3,700 kN, a live load of 1,000 kN, and soil shear strength parameters of cohesion = 100 kPa and friction angle = 30°, is shown in the right column of Table 2.4. For each code, calculation of the required bearing area shows that the Canadian Foundation Engineering Manual is the most conservative (area = 5.2 m^2), and the American National Standard Institute is the least conservative (area = 2.8 m^2).

Rock slope stability - for the rock slope shown in Figure 2.1 and the stability model in Figure 2.2, the LRFD calculation for the (factor of safety) ultimate limit state is as follows:

Base stability model – unfactored: cohesion, c = 25, kPa, friction angle, $\phi°$ = 37°, depth of water in tension crack, Z_w = 2.33 m; Factor of Safety = 1.31.

Based on factors listed in Table 2.4, Resistance factors: $\Phi_c = 0.65$, Φ_φ = 0.8; Load factor: $\gamma_{Zw} = 1.25$.

LRFD model – factored: $c^* = (25 \cdot 0.65) = 16.25$ kPa, $\phi^* = (37 \cdot 0.8) =$ 29.6°, $Z_w^* = (1.25 \cdot 2.33) = 2.91$m; Factor of Safety = 1.02.

Application of these load and resistance factors to the slope stability model reduces the factor of safety from 1.31 to 1.02. The use of a *FS* of 1.02 in design that takes into account uncertainties in shear strength parameters and the groundwater pressure may be acceptable, depending on the possible magnitude of other uncertainties, such as the dimensions of the sliding mass and construction methods, that may also need to be accounted for in design.

For the slope model shown in Figures 2.1 and 2.2, where the sliding surface is a well-defined, persistent, bedding plane, the slope will transition rapidly from stable to collapse such that the ultimate limit states would be essentially identical to the Factor of Safety = 1.0 shear strengths.

2.6.5 Quoted ultimate and serviceability limit states for design

The following comments apply to ultimate and serviceability limit state values that may be used in geotechnical design.

Ultimate limit states (ULS) - the factors of safety listed in Table 2.1, and partial factors quoted in Eurocode 7 (see Section 2.7 below) and used by the United States Departments of Transportation are all for ultimate limit states (ULS) (Orr, 2015) (Transportation Research Board (TRB), 2009). In the Canadian Bridge Design Code, ultimate resistance factors (ϕ_{gu}) values are quoted for bearing, overturing sliding, compression, tension, and lateral limit states, and separate factors are quoted for "low", "typical" and "high" understanding of site conditions depending on the extent of the investigation program. For example:

Retaining systems	Bearing, φ_{gu}	low: 0.45; typical: 0.5; high: 0.6
	Overturning, φ_{gu}	low: 0.45; typical: 0.5; high: 0.55
	Base sliding, φ_{gu}	low: 0.7; typical: 0.8; high: 0.9

Serviceability limit states (SLS): for geotechnical design, serviceability limit states (φ_{gs}) have been applied in highway bridge design for deflection, and for settlement of approach fills where the transition from the approach embankment to the bridge structure is unacceptable for drivability, or where deformations cause unacceptable misalignment, distortion or tilting of the structure. (Canadian Standards Association, 2019). For example:

Retaining systems	Settlement, φ_{gs}	low: 0.7; typical: 0.8; high: 0.9
	Deflection/tilt, φ_{gs}	low: 0.7; typical: 0.8; high: 0.9

That is, these resistance factors recognize that the consequences of inadequate performance due slope instability is likely to be more severe than that for settlement, that is: ($\varphi_{gs} > \varphi_{gu}$) (Fenton, Naghibi, Dundas, Bathurst, & Griffiths, 2016).

The serviceability limit state is particularly applicable to steel and concrete structures where the elastic limits and ultimate strengths of the materials are well defined and the deflection of the structure can be calculated. This situation does not apply as readily to rock and soil where deformation properties of these materials are not as well defined. In summary, for geotechnical design, serviceability limit states may only be applicable to specific, well defined conditions, and that geotechnical structures would usually be designed to the ultimate limit state.

Regarding the ratio of ultimate to serviceability limit states, the ratio ($\varphi_{gs}/\varphi_{gu} \approx 1.1$ to 1.8) for bridge design (Canadian Standards Association, 2019). Also, for rock fall protection structures based on impact energy, full-scale testing shows that the ultimate energy capacity is about twice the serviceability impact capacity. That is, for impact energies less than the serviceability level, rocks are stopped and no maintenance is required. For increasing impact energies, the structure stops the rock but is damaged, and at the ultimate impact energy the rocks are stopped but the damage is so severe that repairs must be made before it can be put back into service. However, the ultimate energy capacity, which is usually determined from full-scale testing, is difficult to determine because tests which destroy or severally damage the structure may have an energy level that is indeterminately greater than the exact ultimate capacity.

2.7 EUROCODE 7

Eurocode 7 is the European standard for geotechnical design that comprises two parts: Part 1, General Rules, and Part 2, Ground Investigation and Testing (CEN 2004. EN 1997-1:2004, 2004); (CEN 2007 EN 1997-2:2007, 2007). Eurocode 7 aims to achieve geotechnical designs with appropriate degrees of reliability using the limit state design method that generally involves calculations with partial factors applied to characteristic parameter values, and their uncertainties (Orr, 2015).

In developing appropriate partial factors for site conditions, Eurocode 7 takes into account three sets of project factors:

i. Geotechnical Category risk (GC1, GC2 and GC3) - see Section 2.7.1,
ii. Different types of ultimate state (EQU, STR, GEO, UPL, HYD) - see Section 2.7.2
iii. Consequence Class (CC1, CC2, CC3) - see Section 2.7.3.

For each of these three sets of project factors, the following site conditions need to be taken into account:

- Ground conditions
- Groundwater situation
- Regional seismicity
- Influence of the environment

The complexity of geotechnical design is also affected by the vulnerability of the proposed structure to its surroundings.

It is noted that Eurocode 7 quotes all partial factors with values >1.0 such that strengths are divided by the partial factors, and the loads are multiplied by the partial factors. Furthermore, the partial factors listed in Eurocode 7 are for ultimate limit states.

2.7.1 Levels of risk and Geotechnical Categories (GC)

The three categories of structures that are classified according to the level of risk are as follows:

a. **GC1** – structures with a low-risk level that are small, relatively simple on ground that is known from comparable local experience to be sufficiently straightforward and does not involve soft, loose or collapsible soil, or loose fill. Other site conditions are no or very low seismic hazard, no excavation below the water table, and no risk of damage to nearby buildings. **Risk – negligible.**

b. **GC2** – conventional-type structures with no exceptional risk, supported by conventional spread, piles or raft foundations on ground that is not difficult and where the ground conditions and properties can be determined from routine investigations and tests. Other site conditions are moderate earthquake hazard, no risk of groundwater damage without warning, and minimal risk of damage to nearby structures. **Risk – not exceptional.**
c. **GC3** – structures with abnormal or high risk that are very large and unusual and on ground that is unusual or exceptionally difficult. Other site conditions are high seismicity areas, high groundwater pressures and high risk of damage to nearby structures. **Risk – high.**

For each GC category, Eurocode 7 describes the scope of the investigations, design procedures, and expertise required for the responsible engineer.

2.7.2 Types of ultimate limit state

Values of partial factors that are applied to representative actions and characteristics resistances are chosen to obtain values that, when used in calculation models, will provide acceptable levels of reliability against both the occurrence of an ultimate limit state and a serviceability limit state. The following five different types of ultimate limit states are identified and defined in Eurocode 7:

a. **EQU** – loss of equilibrium of the structure or the ground, considered as a rigid body in which the strengths of the structural materials and the ground are insignificant in providing resistance.
b. **STR** – internal failure or excessive deformation of the structure or structural elements in which the strength of structural materials is significant in providing resistance.
c. **GEO** – failure or excessive deformation of the ground, in which the strength of the soil or rock is significant in providing resistance.
d. **UPL** – loss of equilibrium of the structure of the ground due to uplift by water pressure (buoyancy) or other vertical actions.
e. **HYD** - hydraulic heave, internal erosion and piping in the ground caused by hydraulic gradients.

The relative significance of, and also the uncertainties in, the actions and resistances of materials differ for these ultimate limit states. Therefore, separate sets of partial factors have been established for each type of limit state.

2.7.3 Consequence Classes (CC) and Reliability Classes (RC)

EN 1990 offers the following two methods of differentiating the reliability of ULS designs:

a. **Consequence Classes** (**CC**) and **Reliability Classes** (**RC**) established by considering the consequences of failure or malfunction of the structure and exposure of the construction works to hazards. The CC classes relate to the loss of human life as well as the economic, social and environmental consequences:
   - CC1/RC1 – low consequence for loss of human life, for example, agricultural buildings;
   - CC2/RC2 – medium consequence for loss of human life, for example, residences and office buildings;
   - CC3/RC3 – high consequence for loss of human life, for example, sports stadia, public buildings.

     The bases of these classes are similar to the bases of the Eurocode 7 Geotechnical Categories listed in Section 2.7.1 above.

     The Reliability Classes are defined with a minimum 50-year target reliability index β for each class as follows: RC1 – $\beta = 3.3$; RC2 – $\beta = 3.8$; RC3 – $\beta = 4.3$. See Chapter 8 for more discussion on Reliability.
b. Introduction of different quality levels for the design, supervision and inspection during execution related to the Reliability Class of the structure.

   For each combination of GC and CC, an appropriate design procedure can be applied.

2.8 CASE STUDIES – QUANTIFICATION OF UNCERTAINTY

Section 1.8 in Chapter 1 introduced four case studies in geotechnical engineering that represent a range of conditions that may be encountered in practice. This section discusses typical Demands and Capacities for the case studies, where these analyses are Task 1 and Task 2 of the overall risk management structure shown in Figure 2.8.

2.8.1 Debris flow containment dam

Construction of a dam to contain debris flows involves first, determining the required containment volume, and second, the dimensions of the dam to contain this volume. These design requirements can be expressed in terms of Demand and Capacity as follows.

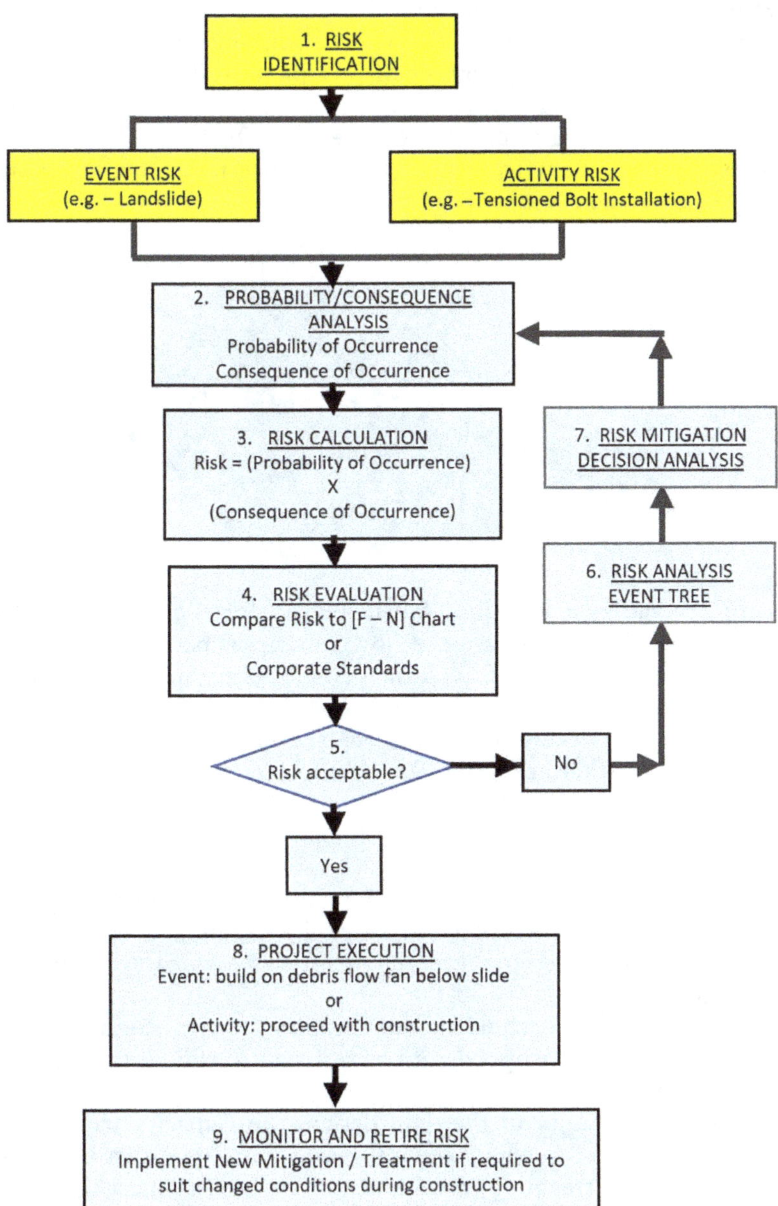

Figure 2.8 Task 1 – Identification of risk for case studies.

Demand - defined by the magnitude-frequency relationship for future events at the site. Definition of this relationship will involve the study of the watershed topography, analysis of precipitation data, examination of the creek bed geology, and investigation of past events evident in test pits excavated in the runout fan (Jakob & Friele, 2010). The estimated volume of future events is likely to have significant uncertainty because none of these design parameters is well defined.

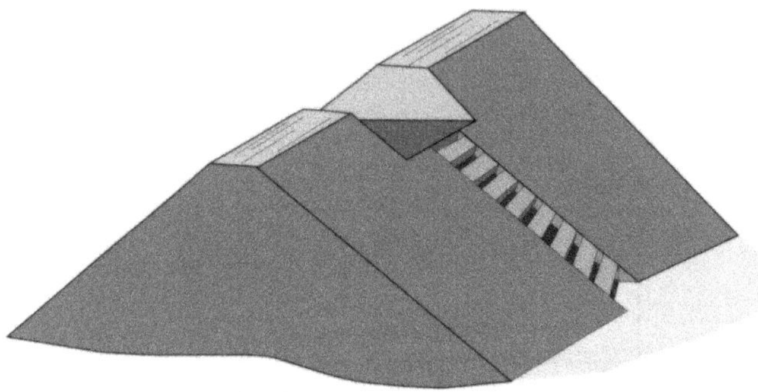

Figure 2.9 shows a probability distribution for the volume of debris flows at a site. That is, the range of flow volumes is between about 2,000 and 30,000 m^3, with a mean volume of 10,000 cu. m. The uncertainty in the volumes is quantified by a lognormal distribution that has no negative values, and is skewed towards the smaller events in the range of 5,000–10,000 m^3 that are more likely than rarer events with volumes of 20,000–30,000 m^3.

Capacity – the design capacity of the dam will be determined by the magnitude of event that is to be contained, taking into account the risk and consequence of damage to structures located downstream on the runout fan. A higher level of protection and lower risk of damage would typically be provided for occupied houses, compared with a low-traffic volume road, for example.

The dam dimensions would be sufficient to contain the design debris flows without being overtopped, and be resistant to sliding-type failure on the foundation.

Regarding uncertainty in the dam design, the Capacity will have low uncertainty compared to the Demand, because the dam can be designed to have a specified strength and volume, and be built to meet these design specifications. A possible source of uncertainty in dam design may be foundation stability if the dam is founded on previous debris deposits.

The design capacity of the containment dam would depend on the acceptable level of risk. As shown in Figure 2.9, a dam with a containment volume of 20,000 m^3 would contain about 95% of all flows, while a volume of 15,000 m^3 would contain about 85% of all flows. Selection of a design capacity would depend on the consequence of the dam being

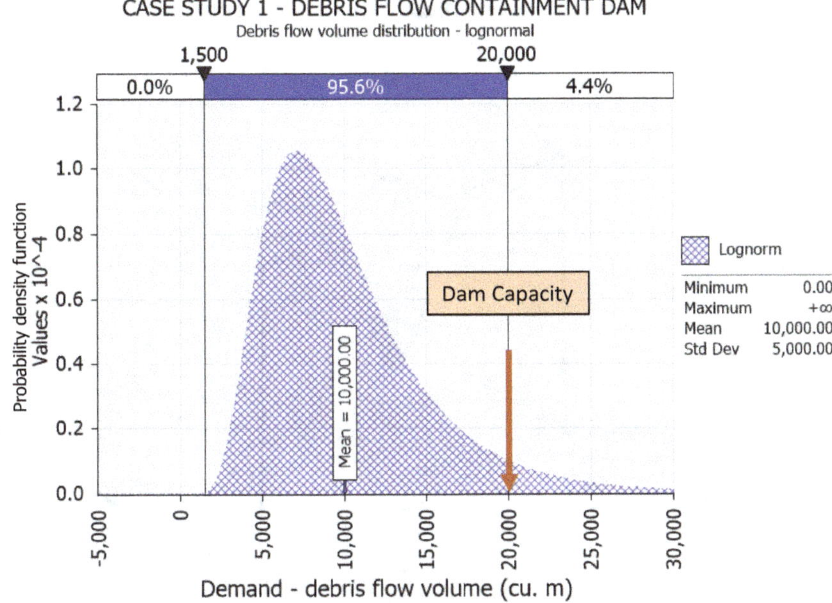

Figure 2.9 Distribution of debris flow volumes.

overtopped, with lower risk being required for high consequences such as loss of life. An additional factor in selecting the capacity is construction cost because the cost of a 30,000 m^3 dam is likely to be significantly more than that for a 20,000 m^3 dam, but would only achieve about a 4% increase in reliability. That is, the extra construction cost may be out of proportion to the reduced risk.

2.8.2 Rock fall hazards

Transportation systems in mountainous terrain can experience rock fall hazards, for which risk management programs may be implemented to limit the consequence of slope instability to their operations. A component of the risk management may be stability analysis and design of reinforcement measures such as installation of tie-back anchors to reduce the risk of slope instability.

The stability of slopes in terms of Capacity and Demand can be demonstrated by wedge-shaped block instability shown in the adjacent sketch, and the simple planar slide model presented in Figure 2.2 that is applicable to the uniformly bedded sandstone shown in Figure 2.1. In reality, most rock masses are more complex than the uniformly bedded sandstone, and they contain discontinuities that have a variety of orientations, spacing and persistence such that uncertainty exists as to the shape and dimensions of potentially unstable blocks, as well as their shear strength values.

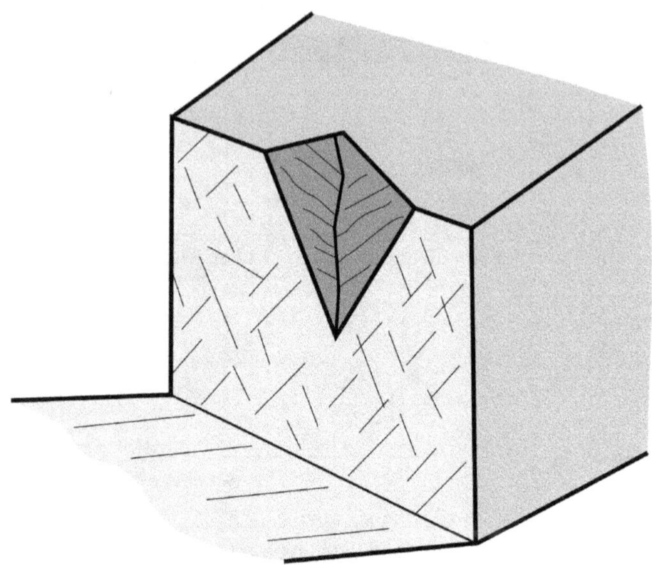

For slope stability analysis, the Capacity and Demand are defined as follows.

Capacity – capacity of a slope to resist failure is provided by the shear strength, defined by cohesion and friction angle, of the soil or rock on the potential sliding plane. If reinforcement, such as tensioned anchors has been installed, the normal and shear components of the anchoring force add to the capacity of the slope.

Demand – demand of a slope is the component of the weight of the sliding mass of the soil or rock acting down the sliding plane, together with any water forces acting in the slope and external forces such as bridge or building loads.

The following comments apply to uncertainty in the Capacity and Demand values.

Regarding **Capacity,** the planar bedding planes in sandstone will have a well-defined shear strength, with a narrow probability distribution, because the cohesion is probably zero, and the friction angle can be measured in the laboratory. In comparison, the rock in the adjacent sketch of the wedge instability contains low persistence joints with variable orientation and spacing such that the sliding surface will comprise both joints and intact rock for which the shear strength will be less well defined and be represented by a wider probability distribution. If tensioned anchors are installed and tested, the value of the support force can be defined by a single value,

Regarding **Demand,** Figure 2.2 shows sliding blocks that have a well-defined shape so that uncertainty in the driving force is low and is represented by a narrow probability distribution. In comparison, the shape and dimensions of the wedges formed in the slope shown in the adjacent sketch will have higher uncertainty, and a wider probability distribution. Water forces in the slope will also be uncertain because they can only be measured at discrete locations, and they vary with time. External dead loads will be well defined, but live loads will have a higher uncertainty.

Stability analyses that take into account uncertainties in the design parameters will have calculated probability distributions for the Capacity and Demand, with the possibility that the two distributions will overlap. As shown in Figure 2.3, overlap of the Capacity and Demand distributions indicates that instability is possible, which is quantified by the probability of failure, *PF*. The probability of failure for this potentially unstable block could be reduced by installing rock bolts that would increase the Capacity, and the slope could be drained to decrease the Demand. Both these measures would decrease the overlap between the Capacity and Demand distributions with a corresponding decrease in the probability of failure.

2.8.3 Tunnel stability

A common method of assessing the stability of rock tunnels and determining support requirements, is to calculate the tunnelling rock mass quality *Q* of the rock in which the tunnel is driven. Figure 2.10 shows the relationship between rock conditions defined by *Q*, tunnel span or height (*m*), importance of the tunnel (ESR), and the corresponding support requirement (Barton, Lien, & Lunde, 1974) (Hoek, Kaiser, & Bawden, 1995). Risk management for tunnel stability can be expressed in terms of Capacity and Demand as follows:

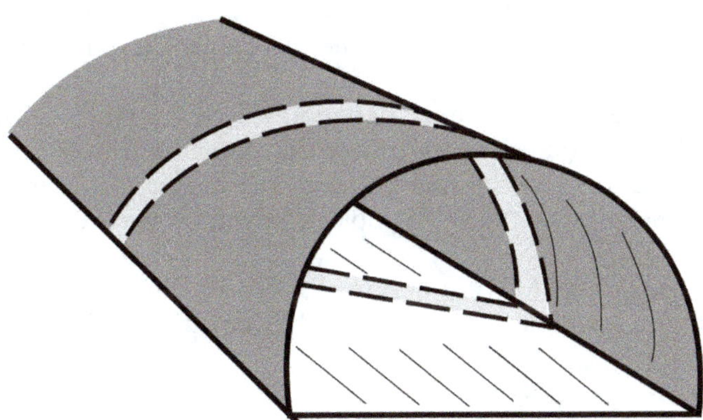

Capacity - installed support required for the tunnel to remain stable. Types of support shown in Figure 2.10 are spot and pattern bolts (*sb* and *B*), unreinforced and steel fibre shotcrete (*S* and *sfr*), steel sets (*RRS*) and

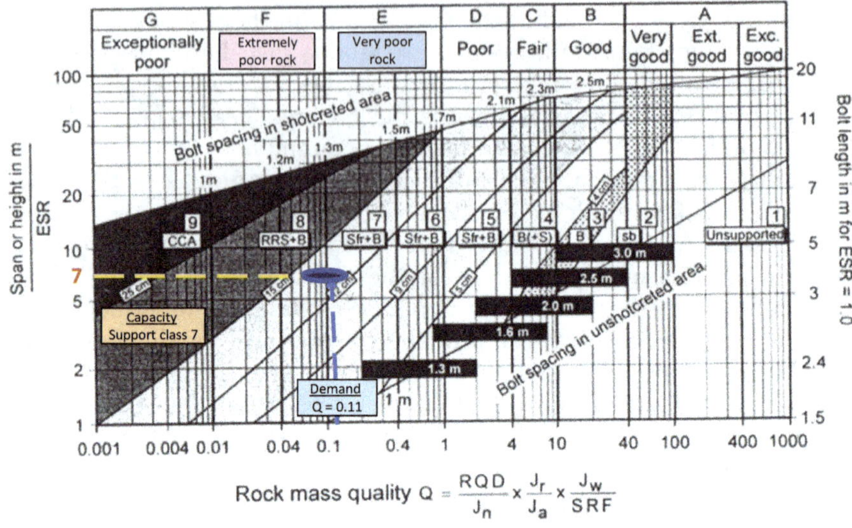

Figure 2.10 Estimated tunnel support categories related to the tunnelling rock mass quality, Q, the span and importance of the structure (ESR); Support Class 7 for Q = 0.11 and (span/ESR = 7).

cast concrete liner (*CCA*). The required support methods can be designed and installed to meet precise specifications such that the support capacity will have little uncertainty.

Demand – tunnel stability is related to the rock mass quality, *in situ* stress, ground pressure and tunnel dimensions, which together define the Demand. Values of all the geological parameters will be uncertain because they are difficult to measure and define, particularly when Q is determined during design and information on the *in situ* geology along the tunnel is very limited.

Risk management in tunnel design involves installing sufficient support for the site conditions such that the tunnel will be stable, and safe for its intended use, taking into account the degree of uncertainty in the Demand. The following is an example of how the Demand uncertainty can be quantified.

Rock mass quality Q is a function of six parameters defining the characteristics of the rock mass that are related by equation (2.21):

$$Q = \frac{RQD}{J_n} x \frac{J_r}{J_a} x \frac{J_w}{SRF} \tag{2.21}$$

where:
RQD = rock quality designation
J_n = joint set number
J_r = joint roughness number

J_a = joint alteration number
J_{aw} = joint water reduction
SRF = Stress Reduction Factor related to loosening of shear zones, and *in situ* stress that may cause squeezing or rock bursts.

Each of these parameters has a range of numerical values defined by the site conditions such that a value for Q can be calculated from equation (2.21). When the Q value is combined with the ratio [tunnel span or height/ ESR] it is possible to estimate the support required for the tunnel as shown in Figure 2.10. For example, for a 9 m wide road tunnel (span/ESR = 9/1.3 = 7), no support would be required for very good rock with a Q value of 100 – Support Class 1, while for very poor rock with a Q value of 0.4, support would be defined by Support Class 6 comprising 9 cm of steel fibre shotcrete (sfr) with 3 m long rock bolts (*B*) installed on a 1.5 m pattern (Grimstad & Barton, 1993). For reference, examples of ESR values are as follows:

- 3 to 5 – temporary mine openings;
- 1.6 – water tunnels for hydropower projects
- 1.3 – minor road and railway tunnels
- 1.0 – major road and railway tunnels
- 0.8 – railway stations, underground nuclear power stations.

The information shown in Figure 2.10 can be used for risk analysis of tunnel stability by quantifying the uncertainty in the rock conditions (Demand). If each of the six parameters in equation (2.21) is assigned a probability distribution, then these can be combined to calculate a resulting probability distribution for Q. For conditions where little information is available on rock conditions at the tunnel depth, the rock parameters could be defined simply by triangular distributions defining the maximum, minimum and most likely values. For example, the RQD would have maximum, minimum and most likely values of 50%, 20% and 40% respectively. The program @Risk would then be used to calculate the corresponding probability distribution for Q as shown in Figure 2.11.

The following is a list of the minimum, most likely and maximum values for the six parameters in equation (2.21). Note that the values for J_n, J_a and SRF have higher values for the poor rock and lower values for the good rock because they are the denominator in equation (2.21).

Q parameter	*Minimum value (poor rock)*	*Most likely value*	*Maximum value (good rock)*
RQD (%)	20	40	50
J_n	12	8	6
J_r	0.5	1.0	1.5
J_a	4	3	2
J_w	0.3	0.4	0.5
SRF	8	5	4

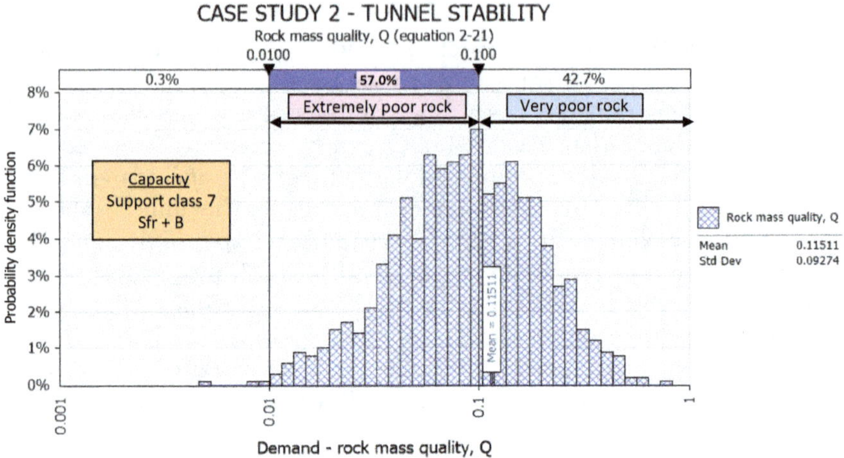

Figure 2.11 Probability distribution for rock quality index Q related to tunnel support requirements as shown in Figure 2.10.

The program @Risk can be used to combine these distributions using equation (2.21) to calculate a probability distribution for *Q* as shown in Figure 2.11. Figure 2.11 shows that the rock quality *Q* values range from 0.01 to 1.0, with 57% probability that the rock will be extremely poor (*Q* = 0.01–0.1) and 42% probability that the rock will be very poor (*Q* = 0.1–1); the mean value of *Q* = 0.12. The risk of instability could be minimized if support is based on extremely poor rock. For a 9 m span, minor road tunnel with an ESR value of 1.3, [(span/ESR ratio) = (9/1.3 ≈ 7)], and the required support is Support Class 7–12 cm thick steel fibre reinforced shotcrete, 4 m long rock bolts installed at a spacing of 1.5 m.

Figure 2.10 shows the rock quality and tunnel dimensions plot on the rock mass quality chart for Support Class 7. Figure 2.10 also shows the uncertainty in the design is represented by a horizontally elongated ellipse representing a range of *Q* values of between 0.06 and 0.2, while the value of the [span/ESR] ratio has little uncertainty.

2.8.4 Dam foundation

The stability of existing concrete dams against sliding on the foundation has sometimes been upgraded to account for seismic ground motions that were not considered in the original design. In these circumstances, a common method of improving stability is to install tensioned rock anchors in holes drilled through the dam into the foundation. The rock anchors are bonded in the foundation with cement grout, and then tensioned against the top of the dam to apply an additional compressive force at the concrete-rock interface.

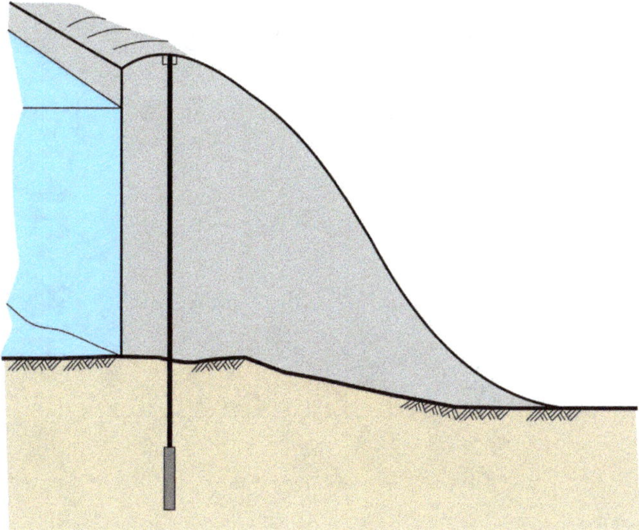

In terms of risk management for these projects, the Capacity and Demand of the dam are as follows:

Capacity – the capacity of a dam is its ability to support external loads such as the contained water, and applicable seismic forces. For a concrete gravity dam, the capacity is determined by the mass of the dam, and its resistance to displacement generated both at the rock-concrete interface and by the shear strength of the foundation rock.

Regarding uncertainty in these parameters, the mass of the dam has low uncertainty, and the strength of the foundation rock will be determined during site investigations and study of the exposed rock during construction. However, the shear strength of the rock-concrete interface, in terms of the cohesion and friction angle, will have a higher degree of uncertainty because these parameters cannot be tested directly.

When tensioned rock anchors are installed to improve resistance against sliding, they will add to the capacity of the dam foundation. The force that the anchors apply will have low uncertainty because the strength of the steel is known and load-elongation tests conducted during installation will verify the load.

Demand – demand forces on a dam are primarily the forces generated by the contained water that comprise both the horizontal force on the face of the dam, and any uplift force in the foundation. The horizontal water force will fluctuate, but the maximum value at full reservoir level will have no uncertainty. However, the uplift force will have some uncertainty because the effect of grouting and drainage on water pressures in the foundation cannot be fully verified.

The demand force with the greatest degree of uncertainty is that generated by seismic ground motions. A common method of modelling seismic ground motions is to use pseudo-static analysis in which a static horizontal

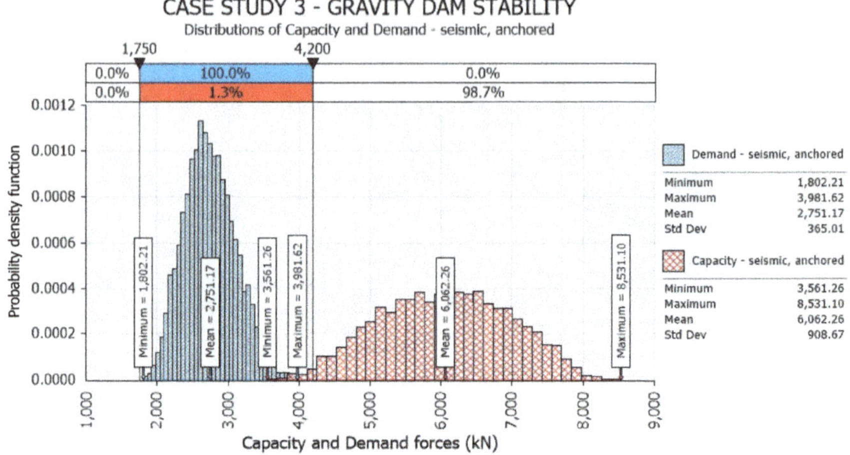

Figure 2.12 Probability distributions for capacity force (μ_C = 6062 kN, σ_C = 909 kN) and demand force (μ_D = 2751 kN, σ_D = 365 kN) on concrete gravity dam with pseudo-static horizontal seismic force, and tie-down anchor force applied.

force, proportional to the earthquake acceleration, is applied to the structure. The magnitude of the pseudo-static force will have a degree of uncertainty because this static force is an approximation of actual ground motions, and the actual magnitude of the earthquake ground motions is unknown.

Figure 2.12 shows the probability distributions for the Capacity and Demand forces, with the higher standard deviation for the Capacity forces (σ_C = 909 kN) compared to the Demand forces (σ_D = 365 kN). The standard deviation in the Capacity is mainly because of uncertainty in the shear strength at the concrete-rock interface, while the standard deviation in the Demand is because of uncertainty in the magnitude of the pseudo-static seismic force. A minor overlap occurs between the two distributions indicating that a very low probability of failure exists.

The mean value of the factor of safety of 2.2 is given by the ratio of the means for Capacity and Demand forces:

$$FS = \frac{Capacity\ force}{Demand\ force} = \frac{6062}{2751} = 2.2$$

The value of plotting distributions such as those shown in Figure 2.12 is that the analyses can examine the influence of uncertainty of all parameters that influence stability.

Chapter 3

Structure of risk management programs

Successful risk management of geotechnical projects requires a well-structured program by a team with experience that is relevant to the project. Risk management can be carried out at all stages of the project so that the initial risks identified during the planning stage of a project can be updated as work progresses and new information becomes available. The up-dating may result in risks being retired if, for example, mitigation eliminates the hazard, or have a higher risk if the ground conditions are more difficult than expected. The usual six stages of geotechnical projects where risk needs to be addressed are as follows:

i. Planning
ii. Investigation
iii. Design
iv. Procurement
v. Construction
vi. Operation and maintenance

The level of effort that is applied to risk management is often a function of project size. For most small projects, with budgets of a few hundred thousand dollars, risk can be assessed by project engineers documenting project conditions and assessing risk qualitatively. However, with increasing project scale and complexity, a more formalized program is required as described in this chapter. As a benchmark on the level of effort that may be implemented, the Washington State Department of Transportation (WSHDOT) requires a full, quantitative risk management program for projects with budgets exceeding $10 million (NCHRP, 2018).

Figure 3.1 shows the basic structure of risk management programs comprising nine tasks from Task 1: Risk Identification through Task 9: Monitor and Retire Risk, with Task 5 being a decision point where it is determined if the risk is acceptable. If the risk is not acceptable, Task 6: mitigation is carried out and Tasks 2, 3, 4 and 5 are repeated until it is decided that the risk is acceptable and Task 8: Project Execution takes place.

DOI: 10.1201/9781003271864-3

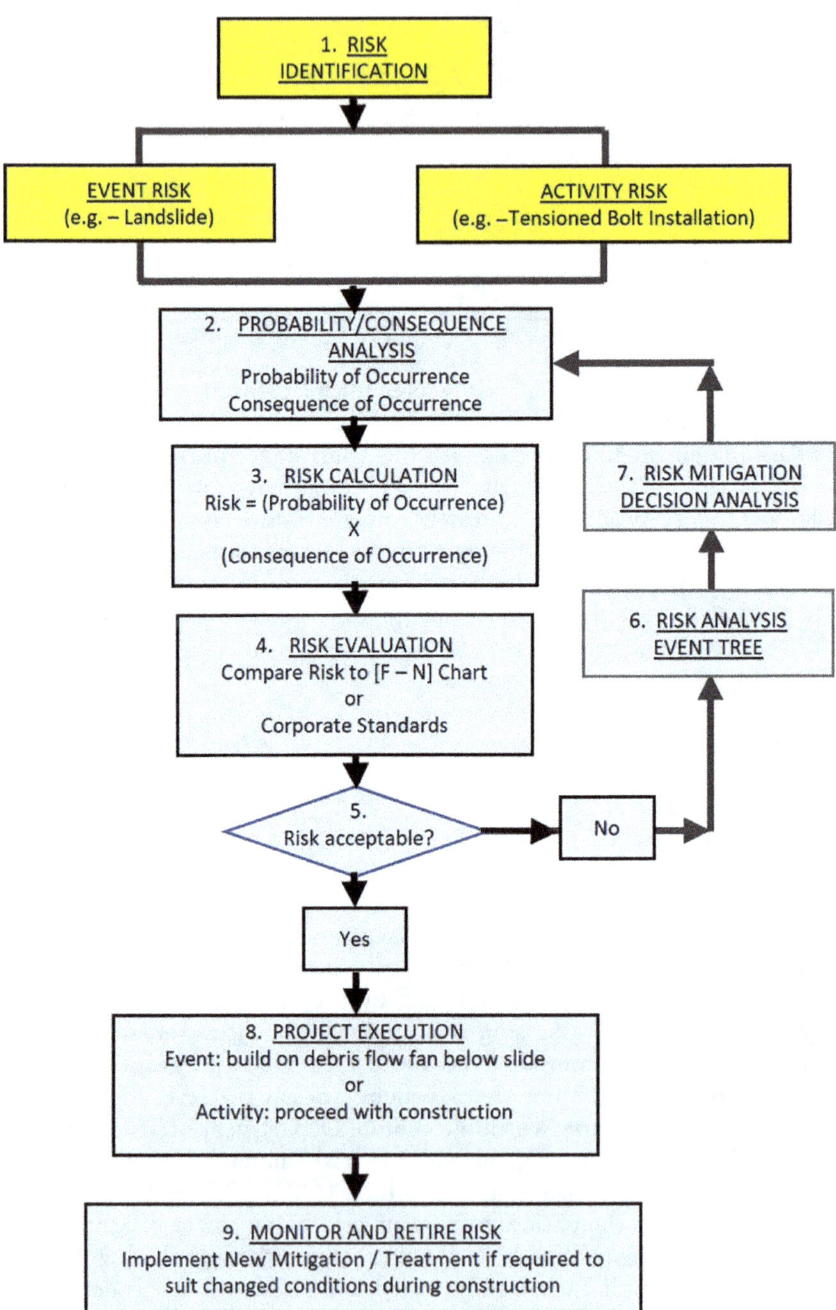

Figure 3.1 Risk management program structure - Tasks 1- risk identification for events and tasks is highlighted.

The first action is Task 1 involving the identification of project risk, both for Events such as landslides and rock falls that can occur repeatably but unpredictably, and for Activities such as installation of anchors that occurs once but procedures are required to limit the risk of unsatisfactory performance.

The risk structure shown in Figure 3.1 is similar to the GeoQ system proposed by GeoDelft, the National Institute for Geotechnical Engineering in the Netherlands (van Staveren, M., 2011). The six tasks in the GeoQ system are:

i. Project information
ii. Risk identification
iii. Risk classification
iv. Risk remediation
v. Risk evaluation
vi. Risk mobilization.

3.1 STAKEHOLDERS

The first step in assessment of project risk is to identify the stakeholders involved. Stakeholders will vary with the type of project and its size, and their interests are also likely to differ depending on their connection to the project. The two major objectives for engaging stakeholders are first, to provide information on the project and gain their acceptance, and second, to obtain background data that may not be readily available to the project team. Regarding background data, for many geotechnical projects much of the relevant part of the structure is buried and details of its design and construction may be forgotten, and the records lost. However, long-term or retired employees may remember vital information that will be valuable for management of risk on the new project.

The following is a list of potential stakeholders on public and private projects, and discussions of their possible understanding, and tolerance, of risk.

3.1.1 Public

On many public infrastructure projects, the public is engaged early in the planning stage in which officials provide information on the project, and elicit comments, concerns and complaints. With respect to risk management, this consultation process will also give the owners feedback on possible opposition to the project, or background information that can be used to improve planning and avoid delays.

In general, the public, who are the users of the project, wants safe facilities such as highways or transit systems that are seen to be well constructed, efficiently operated and have little environmental impact. Construction costs may be of lesser concern unless waste or corruption is evident, or if the project competes with other projects that are perceived to be more urgently needed.

3.1.2 Politicians

The need for public infrastructure projects always exceeds available budgets. Therefore, tradeoffs are required in the selection of projects, and decisions may be driven by election cycles and appeal to voters. With respect to risk management, politician's primary concerns are projects that are on schedule and budget. Regarding technical risk, the main concern is likely to be environmental issues and climate change, with no concern for risk such as stability of excavations.

3.1.3 Regulators

For both public and private projects of almost any size, it is likely that the project must comply with a variety of regulations such as zoning bylaws, building permits, environmental rules and labour laws. The requirements for all these regulations are clearly defined in legal statutes with little room for interpretation or negotiation. Therefore, in terms of risk management for project owners, the concern is the ability to meet the regulation requirements without undue cost, delay or design changes. This can best be achieved by being very familiar with the regulations and allowing sufficient time in the planning stages to satisfy the requirements.

3.1.4 Owners

Project owners may be public organizations such as highway and transit authorities, or private owners such as property developers. While owners have the same risk concerns as the public and politicians regarding budget, schedule and environmental impact, they will also be concerned about technical issues such as encountering subsurface materials that are stronger than expected and require blasting for excavation, or are weaker than expected and are unstable. These types of conditions can result in claims triggered by the Differing Site Conditions (DSC) clause in most contracts. Management of the risk of DSCs involves conducting a thorough investigation program that is reported in the Geotechnical Data Report (GDR), and interpretation of the investigation results as reported in the Geotechnical Baseline Report (GBR). Section 3.7 below discusses these reports in more detail.

Another risk concern for owners is maintenance and operating costs that need to be considered in both design and construction. Control of maintenance costs usually involves tradeoffs between cost and longevity. For example, in reactive soils and rocks, the use of rock anchors fabricated from stainless steel rather than galvanized steel would increase the life of the structure, but at significant upfront cost. Project risk could be optimized by either replacing certain steel components near the end of their useful life or by taking into consideration that many structures are obsolete because they are under capacity or outdated before many of the physical components need replacing.

3.1.5 Engineers

The project engineers will have the central role in identifying, assessing and mitigating risk, with these responsibilities starting with the planning stage of the project and continuing through design, construction and operation. This can be a dynamic process in which risks that are identified in design are monitored during construction such that they are either retired when a component of the project is completed, or are updated if actual site conditions differ significantly from those that were anticipated. At the completion of the project, it may be necessary to again update the risk register to address residual or new risks that should be managed during operations.

For large and complex projects, the engineer may have a person or team of engineers that has full time responsibility for risk management. This arrangement allows for the common situation that risks change as the project proceeds and appropriate adjustments are needed to the original plan (ISO, 2019).

3.1.6 Contractors

The contractor responsible for building the project may have two sets of risks to manage. First, the project risks identified by the other stakeholders such as public acceptance, political, regulatory and design risk, and second, construction risk such as availability of materials, equipment and labour. Regarding the project risk, for some design-build projects, the contractor may be involved early in the project in order to participate in planning and design with the objective of aligning the design with the planned construction methods (NCHRP, 2018). Under these circumstances, the contractor will be an integral part of the project's risk management.

Regarding the contractor's construction risk, hazards and consequences will be documented during bidding and these risks will be built into the bid price, with the risk price discounted to account for mitigation measures that may be implemented. Once the projects start, new risks may arise, or the actual risks may differ from what was anticipated in the initial risk study. Management of these risks may involve making a claim for changed conditions under the DSC clause, for example (see Section 3.1.4) if this can be justified, or implementing additional mitigation.

3.2 STAKEHOLDER ANALYSIS

Having identified potential risk management team members from the list of six possible categories of stakeholders listed in Section 3.1 above, it is also useful to analyze the characteristics of each potential team member to understand their level of interest and support for the project, to see how they might contribute to, or undermine, the process.

It may be useful in the analysis of stakeholder's attitudes to characterize them as follows (Hillson, D., & Simon, P., 2020):

- Attitude is supportive of, or resistant to, project success.
- Power to influence the project is high or low.
- Interest in project high or low

Having identified these basic attitudes, stakeholders can be further analyzed to find who may make positive or negative contributions to the project:

- Saviour – strong, influential supporter
- Friend – high interest in project, but with low influential power
- Sleeping giant – strong supported, but with low level of interest
- Acquaintance – low interest, low influence supporter
- Saboteur – influential with a high level of interest but does not support the project.
- Irritant – high interest in the project but low influence and low support
- Time bomb – influential but with low interest and negative attitude towards the project
- Trip wire – low power, low interest with negative attitude towards the project.

In selecting risk management team members, it is obviously beneficial to select people from the first four categories who support the project. However, people who do not support the project, particularly if they are influential, should be included in the process to some extent because they may be convinced to change their minds. Furthermore, engagement with the non-supporters may contribute to lessening their negative influence on the project.

3.3 EXPERT OPINION AND SUBJECTIVE PROBABILITIES

The stakeholder risk management team may be supplemented with outside expert(s) who can bring special knowledge and experience to the project, and have the advantage of being independent in comparison to stakeholder's closer relationship to the project. The independence of outside experts will enhance the team's ability to provide objective advice and opinions.

Technical issues that the risk team is required to address may not be well defined quantitatively because, for example, they are unique situations that are not found in the historical record, or physical properties that may be poorly understood and the cost of obtaining data could be prohibitive (Baecher, G. B., & Christian, J. T., 2003). For example, a drilling investigation for a deep tunnel may not be feasible because of difficult access to the surface, or economically justified because a few metres length of 75 mm diameter core is unlikely to be representative of the rock in which the tunnel is to be driven.

Obtaining reliable and objective quantification of the uncertainties can be accomplished by having discussions with the team in which the following topics are addressed:

a. Decide on uncertainties for which probabilities need to be assessed.
b. Select a team of experts with balanced spectrum of expertise regarding the uncertainties.
c. Refine with the team the specific uncertainties that need to be addressed.
d. Conduct a short training course with the team on concepts, objectives and methods of assessing subjective probabilities, and the common errors that are made in quantifying probabilities.
e. Have each team member provide subjective probabilities for parameters for which they have experience.
f. Have the team members interact, supported by a facilitator, to discuss their findings with the objective of arriving at a consensus on acceptable probabilities.
g. Document the results of the team's discussions and make them available for the risk management process.

For conditions where the level of uncertainty is high, the risk management team will need to assess probabilities subjectively to determine values of parameters to use in design. Subjective probabilities should be consistent with probability theory (i.e., they should add to 1.0), and be calibrated to real-world values. However, research has shown that people are not well equipped for mentally processing uncertainty, and instead rely on simple mental strategies or rules of thumb (Hogarth, R., 1975). An issue in estimating probabilities is that people rely on complex models while discounting simple observations, and assume that the model results, with their embedded simplifications and assumptions, truly represent field conditions. Another issue is that when people estimate an uncertain quantity, they tend to start with a "best estimate" that is not adjusted sufficiently to reflect the level of uncertainty. A more reliable approach is to first estimate possible maximum and minimum values, and then afterwards select the most likely value that may not be at the midpoint between the maximum and minimum values.

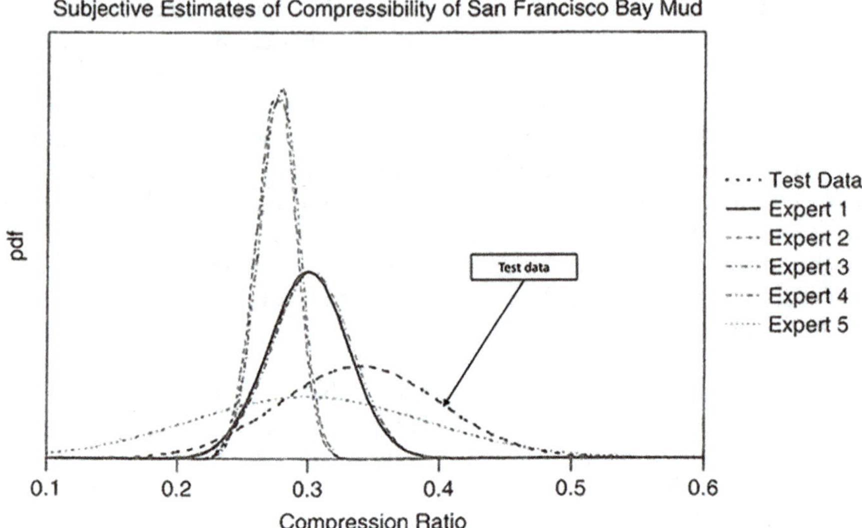

Figure 3.2 Subjective estimates of compressibility of San Francisco Bay mud compared to test results (Baecher, G. B., & Christian, J. T., 2003).

That is, a left skewed result would show that, for example, the shear strength is expected to be close to the minimum value, but the possibility of high strength is not discounted (this issue is also discussed in Section 2.3.3 regarding the use of the "three-sigma" rule to estimate standard deviations).

Reasons for the problems with estimation of subjective probabilities are overconfidence, neglect of base rates and misperceptions of independence (Baecher, G. B., & Christian, J. T., 2003).

First, overconfidence – experts tend to have confidence in their opinions that may be misplaced, particularly with the range of possible values. For example, Figure 3.2 shows subjective estimates of the compressibility of San Francisco Bay mud made by five persons with up to 17 years of experience of working with this material, together with the actual measured compressibility. All five experts underestimated the average compressibility value, but more significantly, four of the five thought that the range of compressibility values was less than the actual range showing overconfidence in their estimation of the parameter uncertainty.

Second, neglect of base rates – for events that are typically rare in nature may be judged by experts to have a reasonably high probability of occurrence based on recent identifiable results within their own experience. This situation is addressed by the "law of small numbers" (Kahneman, D., Slovic, P., & Tversky, A., 1982). That is, an expert may provide a subjective probability based on a small number of samples that are available, and not take into consideration that such a probability is likely to be different from a more reliable value obtained from a large number of samples.

Third, misperceptions of independence – issues of overconfidence and neglect of base rates refer to single events (Kahneman, D., Slovic, P., & Tversky, A., 1982). However, for the combined occurrence of events, the estimated probability may be higher than the probability of each constituent event. For example, the annual probability of a slide from a slope above a highway may be 0.01, and the probability that a vehicle is in the path of the slide and is impacted is 0.0001. Therefore, the annual probability of a vehicle being impacted by a rock fall is: $[(0.01 \cdot 0.0001) = 10^{-6}]$. An estimate made of the annual probability of vehicle impacting a slide may be 10^{-4} because of the difficulty in quantifying the combined effect of two low probabilities rare events.

Another example of uncertainty, related to analysis models, is a study of slope stability calculations using different types of characteristic shear strength values. The analysis results showed factors of safety (*FS*) varying between 0.36 and 1.65. This is a concerning range of calculated *FS* values, particularly where *FS* values are less than 1.0 (Koelewijn, A. R., 2002). However, a hopeful result of this study was that the margin between the minimum and maximum values decreased as more ground data became available from additional field investigations.

3.4 ASSESSMENT OF SUBJECTIVE PROBABILITIES

In situations where subjective probabilities provided by experts and the stakeholder team are an important component of risk management, steps may need to be taken to make these probabilities as reliable as possible. The following is a discussion on three methods of discussing uncertainties and developing a consensus on reliable design parameters.

3.4.1 Brainstorming

Brainstorming involves the risk management team assembling as a group to discuss the design issues, preferably in a structured format in which a facilitator is available to stimulate discussions but not limit ideas. For structured discussions, the facilitator breaks down the issue to be discussed into subtopics and uses prepared prompts to generate ideas on a new topic when one is exhausted. For unstructured discussions, the facilitator starts off a train of thought and everyone is expected to generate ideas, with the pace being adjusted to generate lateral thinking. While working in groups has the benefit of encouraging teamwork, it is found that the negative aspects of brainstorming are (ISO, 2019):

- Fewer ideas are generated by a group compared to people working individually.
- In a group, people's ideas tend to converge rather than diversify.
- The delay in waiting for a turn to speak tends to block ideas.
- People tend to work less hard when working as a group than individually.
- Discussions can be dominated by one or two individuals.
- A skilled facilitator may not be readily available.

Measures to improve the effectiveness of brainstorming are to allow members to work alone for some of the time, and changing team membership occasionally.

3.4.2 Delphi panels

Delphi panels aim to benefit from both the teamwork of brainstorming, and of individual thought. The panel starts by discussing the project in general and the uncertainties in particular, and then each member works alone to develop subjective probabilities for the design parameters for which uncertainty exists. These solutions are then analyzed and possibly combined, with results then shared anonymously between team members who can then reconsider their responses and produce new results. This process is repeated until consensus is reached.

The advantages of Delphi panels are that opinions are anonymous so that unpopular ideas are more likely to be expressed, the process cannot be controlled by dominating personalities, and people have time to make considered responses to the questions.

3.4.3 Structured and semi-structured interviews

In a structured interview, individuals are asked a set of prepared questions, while in a semi-structured interview, the interview allows more freedom for a conversation to explore issues. Questions should be open-ended, simple and cover one issue only, with possible follow-up questions for clarification if required.

The interview process is a means of obtaining in-depth information from individuals in the group, with the responses being kept confidential if required. While the interviews provide considered thought about an issue, as significant limitation is that no interaction occurs between members of the risk management team so the only means of developing consensus is to have the facilitator analyze and summarize the results.

3.5 RISK IDENTIFICATION AND RISK REGISTER

Once the stakeholders have been identified and have been made familiar with the project, the next task is to identify the risks and develop the risk register. The risk register starts with preparing a list of all possible risks on the project, and then refining the list as the project develops.

Identification of the risks may, for large projects, be a formal process in which a team is selected, whose members can bring relevant knowledge and expertise to the project. The function of the team is to identify, assess and where necessary mitigate risk, which will require that a structured procedure for team meetings be set up. To achieve optimum results from the team, a skilled and experienced facilitator is required, with this being particularly important for virtual, electronic meetings. The role of the facilitator is to:

- Organize the team.
- Obtain and distribute project data prior to the meetings.
- Set up an efficient structure and format for the meetings.
- Provoke creative thinking in order to strengthen understanding and generate creative ideas.

When risk identification is carried out early in the project, preferably during planning, this is an opportunity to consider a wide scope of risks, including those thought to have a low probability of occurring. This task should keep in mind the discussions in Section 3.3 that people usually discount the risks of rare events, i.e., very low probability events that may occur on the project and should be incorporated in the risk studies.

Typical steps in refining the risk register are to rank the risks in terms of their likelihood of occurrence, and the consequence that the risk may have on the project. As an aid to the development of the risk register, this section discusses the classification of risk applicable to all projects, and more specifically, lists typical geotechnical risks that have been encountered on highway projects in the United States. For highway geotechnical risks, risks are ranked in terms of their potential consequence on project performance (NCHRP, 2018).

3.5.1 Risk classification (SWOT, etc.)

Identification of risk for a wide range of projects and situations is facilitated using checklists and classifications of risk that can be generated using questionnaires, interviews, structured workshops or remote meetings. It may be helpful in developing lists of risks to apply the following acronyms (ISO, 2019):

- **SWOT - strengths, weaknesses, opportunities, threats** – these risks can apply to internal factors such as damage resulting from a landslide or debris flow, as well as external factors such legal costs and insurance claims resulting from the work.
- **PESTLE, STEEP, STEEPLED** – these acronyms refer to the wide range of possible hazards and consequences that need to be considered on many large-scale projects. The letters in the acronyms refer to the following factors: Political, Economic, Social, Technological, Environmental, Legal, Ethical and Demographic; another factor that should be included in such lists is Regulatory. In the early stages of the risk identification process, all nine factors should be considered – the evaluation may show some factors are of critical importance and need further study, while others are not applicable and can be eliminated from the risk register.

It is likely that for all major, publicly funded infrastructure projects such as highways and transit systems, all nine factors will be of importance. For private projects such as buildings, Economic factors (and schedule) are likely to be the most important, although for most such projects the importance of meeting Environmental, Legal and Regulatory requirements can never be overlooked.

3.5.2 Geotechnical risks on highway projects

As a guideline on possible geotechnical risk that may occur on highway projects, lists have been prepared on risk, and the importance of these risks, on design-build projects (NCHRP, 2018). Table 3.1, which lists 27 geotechnical risks that have been identified by literature searches, case studies and

Table 3.1 List of geotechnical risk factors on highway projects (NCHRP, 2018)

Number	*Risk factor*	*Number*	*Risk factor*
1	Caverns/voids	15	Subsidence (subsurface voids)
2	Chemically reactive ground	16	Existing structures likely to be affected by the work (other than utilities)
3	Liquefaction	17	Contaminated material
4	Karst formations	18	Landslides
5	Rock faults/fragmentation	19	Settlement of adjacent structures
6	Lateral spreading	20	Sensitiveness of public facilities (parks, historic buildings, etc.)
7	Seismic risk	21	Soft compressible soils
8	Underground fabricated debris	22	Groundwater/water table
9	Groundwater infiltration	23	Settlement in general

(*Continued*)

Table 3.1 (Continued) List of geotechnical risk factors on highway projects (NCHRP, 2018)

Number	*Risk factor*	*Number*	*Risk factor*
10	Presence of rock/ boulders	24	Soft clays, organic soils, peat
11	Settlement of bridge approaches	25	Highly compressive soils
12	Eroding/mobile ground conditions	26	Scour of bridge piers
13	Replace *in situ* material with borrowed material	27	Slope instability
14	Unsuitable material		

expert opinion, provides a starting point in risk management to identify potential project risk. While Table 3.1 was developed for highway projects, it is considered the list is equally applicable to linear projects involving excavations and subsurface work such as railways and pipelines.

3.6 RISK ASSESSMENT

Having identified potential risks on the project such as those listed in Table 3.1, the next step is to assess the potential likelihood and consequence of their occurrence. To assist with this assessment, lists have been drawn up showing their frequency of occurrence and importance index. Table 3.2, which lists the ten most frequently encountered geotechnical risks, shows the wide range of conditions that are encountered and the importance of considering all possibilities when assessing risk.

Table 3.3 ranks the 27 risk factors listed in Table 3.1 in terms of their importance based on experience by highway agency employees and contractors on design-build projects. The information contained in Tables 3.1–3.3 can be used both to identify potential risks and to provide a guideline on investigation programs to obtain information on the highest risk conditions.

Table 3.4 is a checklist that can be used to determine which risks are applicable to the project under evaluation, and the level of hazard related to each risk category.

3.7 DOCUMENTING SUBSURFACE CONDITIONS

For geotechnical projects, the greatest source of uncertainty is usually subsurface conditions, and much effort is often expended in investigating these conditions. Investigations can involve geological mapping, drilling, excavating test pits, geophysics, aerial photogrammetry and LiDAR, with the

Table 3.2 Ten most encountered geotechnical risk factors (NCHRP, 2018)

Number	*Risk factor*	*Number*	*Risk factor*
1	Slope instability	6	Landslides
2	Soft clays, organic soils, peat	7	Rock faults/fragmentation
3	Chemically reactive ground	8	Settlement in general
4	Contaminated material	9	Contaminated material
5	Groundwater/water table	10	Karst formations

Table 3.3 Importance index of geotechnical risk factors in design-build projects (NCHRP, 2018)

Risk factor	*Importance index (%)*	*Rank*
Groundwater/water table	43	1
Settlement in general	38	2
Contaminated material	37	3
Soft, compressible soils	36	4
Scour of bridge piers	35	5
Slope stability	35	6
Settlement of bridge approaches	35	7
Highly compressible soils	34	8
Presence of rock/boulders	33	9
Seismic risk	32	10
Soft clays, organic soils, peat	32	11
Settlement of adjacent structure	32	12
Existing structures likely to be impacted by work (other than utilities)	31	13
Unsuitable material	31	14
Landslides	31	15
Sensitiveness of public facilities (parks, historic buildings etc.)	30	16
Underground manmade debris	29	17
Groundwater infiltration	29	18
Replace *in situ* material with borrowed material	28	19
Liquefaction	28	10
Lateral spreading	27	21
Rock faults/fragmentation	26	22
Subsidence (surface voids)	26	23
Karst formations	24	24
Caverns/voids	23	25
Eroding/mobile ground conditions	22	26
Chemically reactive ground	20	27

Table 3.4 Ranked geotechnical risks (NCHRP, 2018)

Category	*Identified risk factors*	*Present*	*Degree of consequence*
High	Landslides		
	Slope instability		
	Contaminated soil		
	Highly compressible soils		
	Settlement of adjacent structure		
	Prediction of subsurface conditions due to inaccessible drilling locations		
	Subsidence due to subsurface voids		
	Soft clays, organic silts, peat		
	Sensitiveness of public consideration (parks, historic buildings etc.)		
	Scour of bridge piers		
Medium	Soft compressible soil		
	Seismic risk		
	Karst formations		
	Caverns/voids		
	Existing structures likely to be impacted (other than utilities)		
	Groundwater/water table		
	Utility conflicts		
	Lateral spreading		
	Liquefaction		
	Rock faults/fragmentation		
Low	Settlement in general		
	Underground manmade debris		
	Settlement of bridge approaches		
	Presence of rock/boulders		
	Erodible/mobile ground conditions		
	Chemically reactive ground		
	Unsuitable ground		
	Groundwater infiltration		
	Replace *in situ* material with borrowed material		

method(s) suited to the site conditions. In reporting the results of the investigations, a common objective is to provide the results in a standard format that is part of the contract documents so that all parties – owners, engineers and contractors – have the same, clearly defined information. Analysis and reporting of geotechnical data have an important role in project risk management because this is a vital opportunity to provide the designers and contractors with information that defines the degree of uncertainty in the data.

Commonly used documents that define project conditions are the *Geotechnical Data Report* and the *Geotechnical Baseline Report* as discussed below (Essex, 2007). Although these reports were originally developed for the tunnelling industry where risks are high and cost overruns common, they can be readily applied to any geotechnical project.

3.7.1 Geotechnical data report (GDR)

The GDR lists the results of the investigation program, including fieldwork and laboratory testing, without interpretation. The scope of the GDR will show where data is available, and where it is missing. For example, drilling may not be possible in areas where existing buildings are located, or ground water information may only be collected for a short period so that seasonal variations of the water table are unknown.

In many contracts, the contractors are given the opportunity to conduct their own investigation program to supplement information provided in the contract documents. This action would constitute risk management on the part of the contractor to provide information specific to the planned construction method and bid price.

3.7.2 Geotechnical baseline report (GBR)

The GBR is a construction-focused document, the principal purpose of which is to set clear, realistic baselines for conditions anticipated to be encountered during subsurface construction. The document provides all bidders with a single contractual interpretation that can be relied upon in preparing their bids (Essex, 2007). The baseline values are the best estimate of the engineer's interpretation of site conditions, and the key objectives of the GBR include:

- Presentation of geotechnical and construction considerations that formed the basis of design for the subsurface components and for specific requirements that may be included in the specifications.
- Enhancement of the Contractor's understanding of the key project constraints, and important requirements in the contract plans and specifications that need to be identified and addressed during bid preparation and construction.
- Assistance to the Contractor of Design-Build (DB) team in evaluating the requirements for excavating and supporting the ground.
- Guidance to the Owner in administering the contract and monitoring performance during construction.

The GBR is more than a collection of baselines. The report is the primary contractual interpretation of subsurface conditions, and the report discusses these conditions in enough detail to accurately communicate these conditions to bidders, and explains the rational for the baselines. The GBR allocates risk depending on how the baselines are defined. The report is also a risk management tool because it can address the resolution of circumstances outside the baselines.

Note that the GBR replaces the Geotechnical Interpretive Report (GIR), and it is not necessary or advisable to prepare both a GBR and a GIR to make sure that no conflict occurs between the two documents.

3.7.3 Geotechnical memoranda for design

During the investigation and design phase of a project, a series of memoranda may be prepared describing the status of the geotechnical information, as well as interpretations of this data and assumptions on how the data may be used in construction. These memoranda can be used as a basis for internal discussion between project team members, with the objectives of progressing the design and eventually producing a consensus GBR document. Once the GBR is produced, the internal memoranda are not a component of the contract documents because the GDR and the GBR are the only and primary documents that describe and define the geotechnical conditions for the project.

3.7.4 Differing site conditions (DSC)

The primary purpose of the baseline statements in the GBR is to assist in the administration of the Differing Site Conditions (DSC) clause. The DSC clause was developed to take at least some of the gamble on subsurface conditions out of the bidding process, and thereby reduce bid prices. Without the relief on the DC'S clause, the Owner would assign all the risk to the Contractor, and would thus pay all the Contractor's contingency cost for adverse conditions, whether or not adverse conditions were encountered. The DSC clause was developed to avoid these unnecessary costs, and remove part of the risk from the Contractor (Essex, 2007).

The DSC clause, together with the baselines in the GBR, were intended to define geotechnical conditions for all parties in the contract, and to avoid the situation where the contact contained disclaimers that bidders should not rely on boring logs and other information obtained during design, and that they should make their own site investigations. However, in several court cases it has been established that bidders have the right to rely on this information, despite the information being disclaimed and not included in the contract documents.

The following is a standard DCS clause included in many United States federal contracts that involve subsurface construction.

DIFFERING SITE CONDITIONS (APRIL 1984)

a. The contractor shall promptly, and before such conditions are disturbed, give a written notice to the Contracting Officer of: (1) subsurface or latent physical conditions at the site that differ materially from those indicated in the contract, or (2) unknown physical conditions at the site, of an unusual nature, which differ materially from those normally encountered and generally recognized as inhering in work of the character provided for in the contract.
b. The Contracting Office shall investigate the site conditions promptly after receiving the notice. If the conditions do materially so differ and cause an increase or decrease in the Contractors' cost of, or time required for, performing any part of the work under this contract, whether or not changed as a result of the conditions, an equitable adjustment shall be made under this clause and the contract altered in writing accordingly.

The function of the DCS clause is twofold. First, it relieves the Contractor of assuming the risk of encountering conditions differing materially (i.e., in a significant, meaningful way) from those indicated or ordinarily encountered. Second, it provides a remedy under the construction contract, to handle the matter as an item of contract administration.

It is noted that courts may have made more recent interpretations of the DSC clause and these judgements should be studied, if appropriate.

3.8 CASE STUDIES – STAKEHOLDERS AND RISK IDENTIFICATION

As a follow-up to the discussion in Section 3.1 on Stakeholders, and the Identification of Risk in Section 3.5, these issues are illustrated below in four case studies.

3.8.1 Debris flow containment dam – event risk

A gravity dam structure is planned to contain debris flows in a creek that flows in an incised gully down a steep mountain slope, and then through a residential area that has been constructed on the runout fan. In the event of a substantial debris flow, the houses are vulnerable to damage and the residents are at risk from injury and death. The planned remediation for the site is to construct a containment dam at the apex of the fan where the creek is still confined in the gully, assuming that it is not feasible or economic to relocate the houses.

Debris flow hazard is an event-type risk because debris flows occur periodically, with varying magnitude, and uncertainty exists as to the magnitude-frequency relationship for these events.

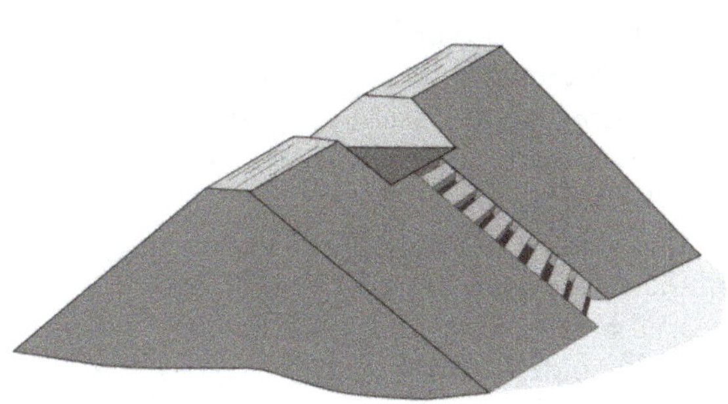

Planning for construction of the dam involves identification of both stakeholders and the project risks.

Stakeholders – local municipality that commissioned the barrier; higher levels of government that provide partial funding; expert in debris flows employed to define the frequency-magnitude relationship; engineering company with specialist expertise to design the dam; consultant carrying out environmental impact study; prime contractor with sub-contractors to construct the dam. The residents will be informed and consulted about the project, but are unlikely to be part of the risk management team.

Risk identification – the primary risks that are of concern for the residents are that the dam will be overtopped, or might fail, resulting in flooding on the runout fan. The design engineer will have responsibility for convincing all stakeholders that the dam will provide the required level of protection based on frequency-magnitude data that will determine the capacity of the dam. The other major risk that will be of concern to the funding agencies will be the construction cost estimate and the risk of cost overruns. Possible uncertainties in the costing are the strength of the dam foundation, and the availability of suitable construction materials. The debris flow expert will provide assistance to the municipality that probably has limited experience with this type of project.

3.8.2 Rock fall hazard – event risk

A 60 km length of a railroad is located in steep, mountainous terrain in a climate with high rainfall and cold winters. The hazard is that these conditions cause occasional rock falls that can result in delays to traffic and in some cases, derailments, damage to equipment and injury to personnel. The economic consequence of these events is often significant because this is the mainline with no alternative routes onto which traffic can be diverted. Risk management for the railroad involves preventing rock falls from reaching the track by either stabilizing the rock cuts, or constructing protection structures such as Attenuator fences or reinforced concrete rock sheds (Wyllie, 2014).

The slope stabilization program is an event type risk because slope instability occurs periodically and at an uncertain frequency. The stakeholders for the stabilization program and the risks for its implementation are discussed below.

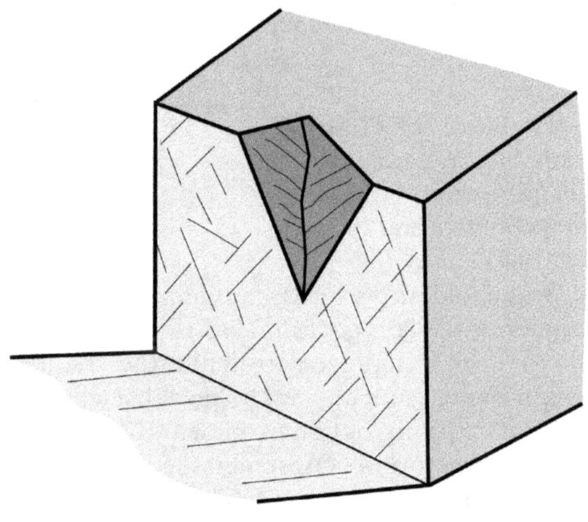

Stakeholders – railroad company that operates the trains; government agency responsible for transportation safety; train operators (engineers); engineering group responsible for analyzing slope stability and design of remedial measures; speciality contractor with experience in working on steep rock slopes.

Risk identification – risk to railroad operations is related to the uncertainties in the locations of rock falls, the frequency of these falls and their consequence. Furthermore, because of the difficult access to the rock slopes and the need to maintain traffic with minimal shutdowns, stabilization work can only be carried out in limited locations each year. This means that it is important to use a system to quantify stability conditions so that slopes can be ranked in terms of the likelihood of instability and that the stabilization work is prioritized. It is usually necessary to repeat this slope stability evaluation procedure annually because conditions deteriorate over time, particularly during the winter, and a revised priority list for stabilization work may be required (Wyllie, Brummund, & McCammon, 1979).

A second risk is related to the longevity of stabilization work. For example, removal by hand of loose rock with a steel pry bar (scaling) is a quick and inexpensive method of improving stability that may be appropriate in an emergency. However, after scaling, the rock will continue to degrade, and the slope may become hazardous again in a few years. Alternative stabilization measures with greater longevity would be installation of rock bolts, particularly if they are galvanized, and the application of shotcrete, particularly if it is tied to the heads of the rock bolts to reduce the risk of delamination from the rock surface. A method of attaching shotcrete to the bolts is to reinforce the shotcrete with welded wire mesh that is placed under the plates on the rock bolts. For this type of high-quality stabilization, the longevity of the support may be 20–50 years depending on the weather and rock quality.

3.8.3 Tunnel stability – activity risk

A two-lane highway tunnel, with dimensions of 12 m wide and 6 m high will be driven by drill and blast methods. The tunnel is located in complex volcanic terrain where details of the geology are uncertain because difficult access limits investigation drilling. The geological uncertainties relate to the possibility of intersecting a major fault or contact containing weak rock as well as high-pressure ground water. Another possibility is that some formations may contain acidic rock that needs treatment before disposal.

Activity-type risk related to the construction of the tunnel, for which the stakeholders and risk identification are discussed below.

Stakeholders – highway authority in whose jurisdiction the tunnel is located; engineering company responsible for investigating the geology and designing the tunnel support; construction company conducting the drill and blast operation. The public would be informed about the project but would not be a part of the risk management team.

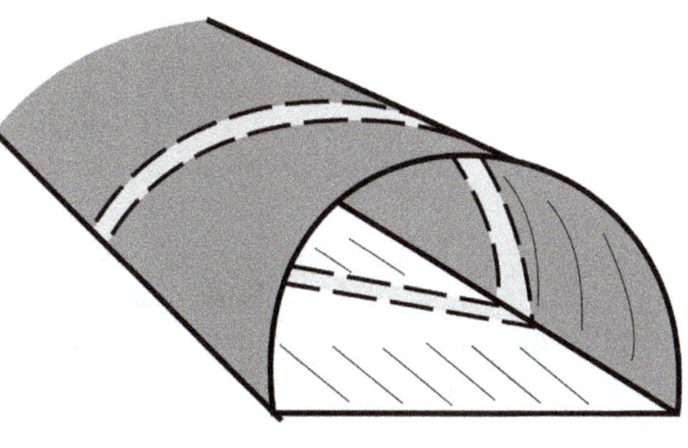

Risk identification – major risk is unanticipated intersection of a zone of weak rock, particularly if it contains large volumes of water. Consequences of this condition could be sudden collapse of the tunnel possibly resulting in loss of life and damage to equipment, as well as schedule delays and cost overruns. Mitigation measure would be to drill probe holes ahead of the face to identify poor rock conditions so that the ground could be reinforced ahead of the face, using grouting and forepoling, for example. Regarding acidic rock, the mitigation measure would be to develop a separate waste rock disposal area where acidic rock that is encountered could be treated or contained.

3.8.4 Dam foundation – activity risk

An existing gravity dam, which is presently stable, will be up-graded to improve its stability in the event of seismic ground motions. The upgrade will involve installation of a series of multi-strand cable anchors, installed in holes drilled vertically through the dam and anchored in the rock foundation. The anchors will be tensioned against the crest of the dam in order to increase the normal stress in the base of the dam at the concrete rock interface. The increased normal stress will improve the stability of the dam against sliding. Successful installation of the anchors requires drilling large diameter, precisely aligned holes, and bonding the distal ends of the anchors in sound rock with no loss of grout in the bond zone.

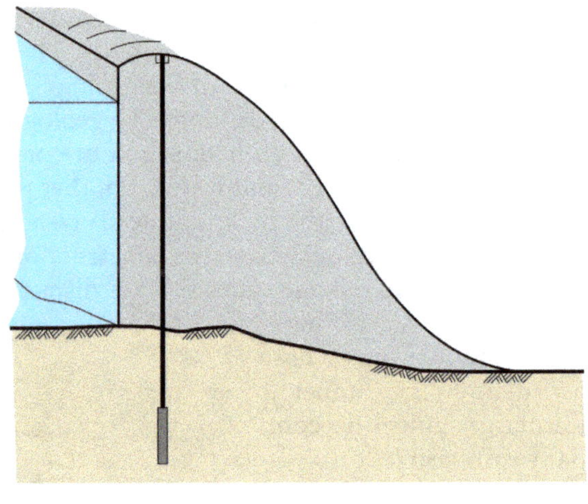

This project is an Activity-type risk related to the installation of the anchors, for which the stakeholders and the risks are discussed below.

Stakeholders – utility company that owns the dam; government regulatory agency which ensures that all dams meet current safety standards; engineering company that designs the anchor system; contractor who will install and tension the anchors. For this type of project, the public is unlikely to be involved.

Risk identification – while installation of tensioned anchors is simple in principle, details of the installation are important because of the very high, concentrated shear stress that is applied in the bond zone at the rock-grout interface. Project risk is related to maintaining alignment of the drill holes within the body of the dam, and intersecting rock in the bond zone that is both strong enough to sustain the applied stresses, and has low hydraulic conductivity so that loss of grout into the rock does not occur. The value of the load applied to the strands and then locked-off in the anchor, is verified by conducting load-deformation tests and ensuring that standard acceptance criteria are met (Post Tensioning Institute, 2014). A significant project risk is failure of the bond zone; the bond is difficult to improve once slippage has occurred, and it may be necessary to replace the anchor at significant extra cost.

Chapter 4

Risk analysis

Once the stakeholders and the project risk team have identified all potential risks as discussed in Chapter 3, the next step is to analyze the risk as described in this chapter. This analysis involves separate examination of the hazard(s) and the consequence(s) as shown in Task 2 in Figure 4-1, and then combining them to determine the risk according to the relationship:

$$\text{Risk} = (\text{probability of occurrence}) \cdot (\text{consequence of occurrence}) \qquad (4.1)$$

The calculated risk can then be compared to acceptable risk levels that are defined by either society or corporate policy, as discussed in Chapter 5. If the risk level is unacceptable, mitigation measures are required to either reduce the probability of occurrence of the event or the magnitude of its consequence, if the project is to proceed, as discussed in Chapter 6.

In preparing risk analyses for complex projects where a thorough, quantitative analysis is justified, the study would include all risks and not just geotechnical issues. An example of non-geotechnical risks could be public infrastructure projects where public acceptance and use of the facility would be a vital measure of its success. Issues that may influence the acceptance of a project are safety concerns such as falls of concrete ceiling panels in the Boston Big Dig highway tunnels (New York Times, 2007), or high cost for usage such as traffic tolling charges.

4.1 EVENT RISK AND ACTIVITY RISK

Two types of risk that may occur on projects are Event risk and Activity risk, for each of which the analysis approach is slightly different (see Section 1.4). For the four case studies described in this book, the debris flows and rock falls are Event-type risks, and the tunnel stability and dam anchoring projects are Activity-type risks. For both types of risk, the consequences of unsatisfactory performance can be financial losses, schedule delays, environmental and regulatory consequences, or injury to persons (Christian,; Beacher, 2011).

 DOI: 10.1201/9781003271864-4

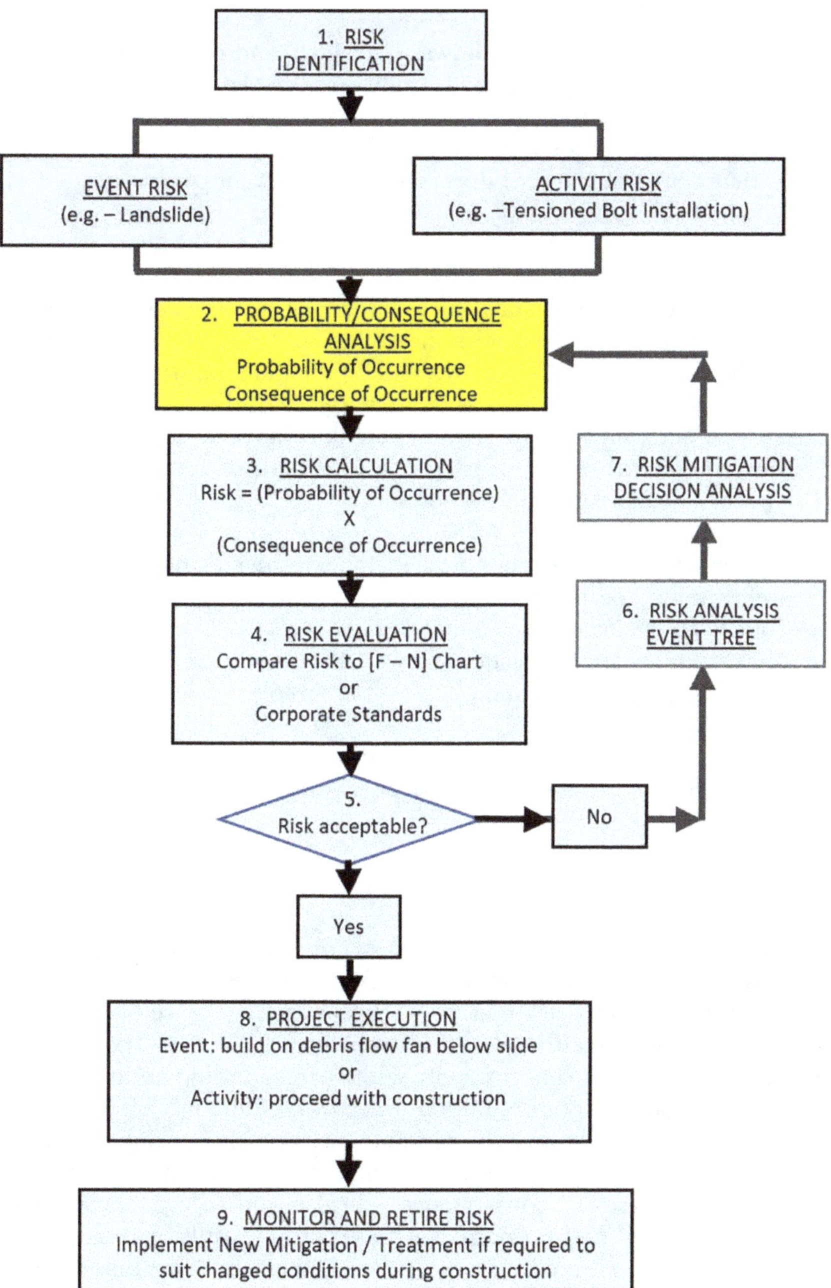

Figure 4.1 Risk management structure - Task 2 probability and consequence of hazard occurrence.

4.1.1 Event risk

Events are risks such as debris flows, landslides and rock falls that occur repeatedly but unpredictably, and study of records of events that have occurred in the past can be used to estimate the annual frequency of future events. The annual frequencies are usually expressed as probability distributions that define the level of uncertainty in the event occurrence.

4.1.2 Activity risk

Activities that occur only once, such as driving a tunnel through a fault zone, can also have risks such as instability and possibly collapse of the tunnel. The risk of instability can depend on the uncertainty in the geology, as well as execution issues such as the experience of the contractor and the suitability of equipment being used. The risk of adverse consequences can be estimated qualitatively by scoring risk on a scale of 1 to 5. Descriptions of the 1 and 5 scale risks are as follows:

- **1 – very low/rare/improbable** – if it is considered that the geology is well known and the contractor has the experience and equipment to cope with the anticipated conditions, to
- 5 – **very high/almost certain/very sure** – if the level of uncertainty in the tunneling geology and operation is severe.

These qualitative descriptions and their numeric values, based on the Likert scoring system, are listed on Table 4.1 in Section 4.4.

4.2 RISK MATRIX

Analysis of the probability of hazard occurrence, the consequences of these hazards, and calculation of the risk of the event or activity, can be shown graphically on a risk matrix (Figure 4.2). On Figure 4.2, the vertical axis shows the likelihood of occurrence expressed as six categories from almost certain to very unlikely, with corresponding annual frequencies from >0.9 (annually) to <0.0001 (once in 10,000 years). The horizontal axis of the matrix shows the possible outcome or consequence of the event expressed as six categories from incidental to catastrophic, and for four potential classes of consequence – health and safety, environment, social and cultural, and economic. An event may result in consequences that influence one or more classes of consequence and for these circumstances, it would be usual to select the most severe consequence for the risk calculation.

Having selected the appropriate likelihood of an event occurrence, and its consequence, the matrix will show the risk level of the event, ranging

Multi-Hazard Risk Evaluation Matrix (SAMPLE)
For the Qualitative Assessment of Natural Hazards

Risk Evaluation and Response		
VH	Very High	Risk is imminent; short-term risk reduction required; long-term risk reduction plan must be developed and implemented.
H	High	Risk is unacceptable; long-term risk reduction plan must be developed and implemented in a reasonable time frame. Planning should begin immediately.
M	Moderate	Risk may be tolerable; more detailed review required; reduce risk to As Low As Reasonably Practicable (ALARP).
L	Low	Risk is tolerable; continue to monitor and reduce risk to As Low As Reasonably Practicable (ALARP).
VL	Very Low	Risk is broadly acceptable; no further review or risk reduction required.

Likelihood Descriptions	Partial Risk (Annual Probability) Indices		Probability Range						
Event typically occurs at least once per year.	F	Almost Certain	>0.9	M	H	H	VH	VH	VH
Event typically occurs every few years.	E	Very Likely	0.1 to 0.9	L	M	H	H	VH	VH
Event expected to occur every 10 years to 100 years.	D	Likely	0.01 to 0.1	L	L	M	H	H	VH
Event expected to occur every 100 years to 1,000 years.	C	Possible	0.001 to 0.01	VL	L	L	M	H	H
Event expected to occur every 1,000 years to 10,000 years.	B	Unlikely	0.0001. to 0.001	VL	VL	L	L	M	H
Event is possible but expected to occur less than once every 10,000 years.	A	Very Unlikely	<0.0001	VL	VL	VL	L	L	M
			Indices	1 Incidental	2 Minor	3 Moderate	4 Major	5 Severe	6 Catastrophic
Description of expected negative outcome (Consequence)			Health & Safety	No impact	Slight Impact: recoverable within days	Minor injury or personal hardship; recoverable within days or weeks	Serious injury or personal hardship; recoverable within weeks or months	Fatality or serious personal long-term hardship	Multiple fatalities
			Environment	Insignificant	Localized Short-Term Impact: recovery within days or weeks	Localized long-term impact; recoverable within weeks or months	Widespread long-term impact; recoverable within months or years	Widespread impact; not recoverable within the lifetime of the project	Irreparable loss of a species
			Social & Cultural	Negligible Impact	Slight Impact to Social and Cultural Values: recoverable within days or weeks	Moderate impact to social & cultural values; recoverable within weeks or months	Significant impact to social & cultural values; recoverable within months or years	Partial loss of social & cultural values; not recoverable within the lifetime of the project	Complete loss of social & cultural values
			Economic	Negligible: No business interruption	<\$10,000 business interruption loss or damage to public or private property	<\$100,000 business interruption loss or damage to public or private property	<\$1M business interruption loss or damage to public or private property	<\$10M business interruption loss or damage to public or private property	>\$10M business interruption loss or damage to public or private property

Figure 4.2 Sample qualitative risk evaluation matrix for landslide events (Porter, M. & Morgenstern, N., 2013).

from very low (VL) to very high (VH). The upper text box on Figure 4.2 provides an evaluation for each of the five levels of risks that define the appropriate mitigation action that may be considered.

The matrix shown in Figure 4.2 is applicable to natural hazards such as debris flows and landslides, but can readily be adapted to other types of projects, including activities.

4.3 QUALITATIVE AND QUANTITATIVE ANALYSIS OF RISK

The decision to use qualitative or quantitative risk analysis, or both, depends on the size and complexity of the project, and its stage of development. As a guideline, Washington State Department of Transportation carries out full quantitative risk analysis for highway projects with a cost greater than $10 million, but does not discourage such analyses for projects with costs less than $10 million.

The other factor influencing the level of detail of risk analysis is the project's stage of development. That is, during conceptual planning and design, a qualitative risk assessment would be appropriate in which the major risks are identified and ranked, because at this time only preliminary geotechnical data would be available, and accurate estimates of cost and schedule consequence may not be feasible (Figure 4.3). However, a quantitative risk analysis will require collection of data on which to base the analysis, and for small projects the cost of obtaining geotechnical data, such as drilling and groundwater studies for example, may not be justified.

The GeoQ risk management system developed by GeoDelft, the National Institute for Geo-Engineering in the Netherlands, has a similar breakdown for the stages of a project related to management of risk, as follows (van Staveren, M., 2011):

i. Feasibility
ii. Pre-design
iii. Design
iv. Contracting
v. Construction
vi. Maintenance

The progression of projects through these six phases allows knowledge of geotechnical conditions to increase as investigation programs are carried out, and actual conditions become evident during construction. With the

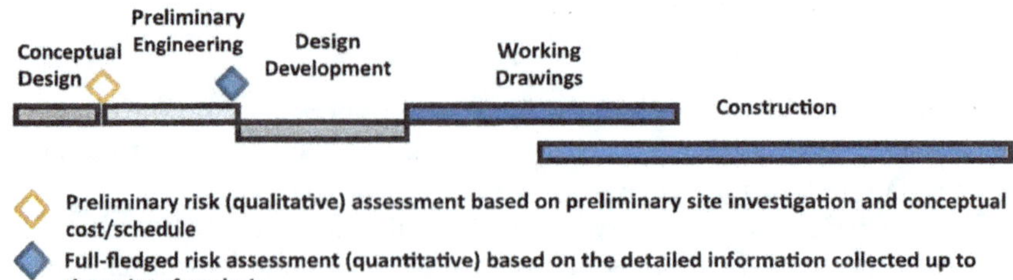

Figure 4.3 Stages of project development and timing of risk analyses (NCHRP, 2018).

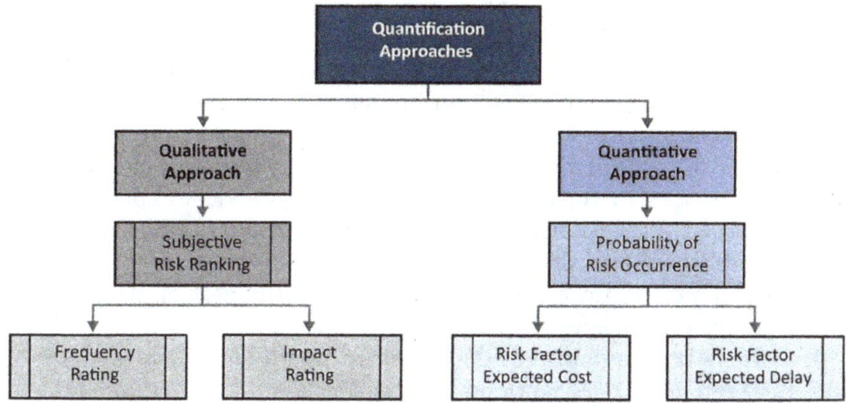

Figure 4.4 Approaches for qualitative and quantitative risk analysis.

increase in knowledge of ground conditions, a corresponding decrease in uncertainty in ground risk would usually occur. The GeoQ process assists in carrying out appropriate investigations to reduce the ground risk to acceptable levels.

Figure 4.4 shows how information used in the qualitative and quantitative types of risk analysis differ as the project develops and more data becomes available. That is, during conceptual design, the frequency of occurrence of events and their consequence will be based on opinions and experience, while during detailed design, expected costs and delay times can probably be defined numerically.

4.4 QUALITATIVE RISK ANALYSIS

Despite the scarcity of information in the early stages of a project, risk analysis is still a useful exercise to help identify potentially high risk conditions that should be investigated so that they are better defined for final design. Often, subsurface conditions have the greatest uncertainty of all design parameters, and decisions have to be made on the extent of investigation programs that should be carried out to reduce uncertainty to an acceptable level. For highway projects it is found that, while geotechnical issues may dominate the risk factors, they are not the most important factor in deciding on the project delivery method. For example, selection of Design-Build as the preferred contracting method is more often driven by schedule and political issues, rather than geotechnical conditions. However, conditions such as active seismic zone and liquefaction potential, or a strong possibility of encountering contaminated ground are cited as reasons for eliminating the Design-Build alternative (NCHRP, 2018).

In the early stages of projects when limited definitive data on risk is available, it is possible to rank hazard frequency and consequence using a Likert rating scale of 1 to 5 where 1 is, for example, a very low frequency of occurrence and 5 is a very high frequency. Initially, the rankings will have to be based on judgement, which will be up-dated as more information is gathered on the project. The selection of hazard frequency and consequence will involve discussions by the risk management team, and possibly more formal evaluations by the team together with independent experts selected on the basis of their particular knowledge of site conditions. The formal process of drawing up risk scores may include the involvement of a facilitator who conducts brainstorming sessions, Delphi panels or structured or non-structured interviews as discussed in Section 3.4.

To assist in interpretation of the Likert scales, Table 4.1 shows relationships between verbal descriptions of the ratings and approximate numeric percentages of occurrence.

Once each of the hazards and consequences have been rated subjectively, they can be assigned Likert scores, such as high likelihood of occurrence (rating 4) and minor consequence (rating 2) for a risk score of: [4·2=8]. The word "likelihood" is often used at the qualitative stage of projects to express the possibility of occurrence of an event because it is indicative of the uncertainty in occurrence, compared with the word "probability" that implies that a (precise) numeric value has been determined for the possibility of the occurrence.

To assist in assessing the relative importance of risks that have been identified for a project, guidance may be provided by consulting the list of geotechnical risks in Tables 3.1–3.4 that identify and rank risks for highway

Table 4.1 Qualitative terms (Likert scales) and transition to numeric judgement

Likert score	*Probability (Likelihood)*	*Synonyms*		*Approximate % of occurrence*
5	Very high	Almost certain	Very sure	>90
4	High	Likely	Reasonably sure	80
3	Medium	Possible	Maybe	50
2	Low	Unlikely	Seldom	20
1	Very low	Rare	Improbable	<10
Likert score	*Consequence (consequence)*	*Synonyms*		*Approximate % of occurrence*
5	Extreme	Very critical	Catastrophic	>10
4	Major	Critical	Severe	8
3	Moderate	Moderate	Major	4
2	Minor	Mild	Minor	2
1	Minimal	Very mild	Incidental	<1

projects in the United States. The information contained in these tables should have reasonably wide applicability because of the diverse geology of the United States. For projects outside the United States, consideration should be given to differing geology or climate at the local site, such as deep weathering in tropical areas.

The likelihood and consequence ratings can then be combined to calculate the risk according to the usual relationship that defines risk:

$$\text{Risk} = (\text{likelihood rating}) \cdot (\text{consequence rating}) \tag{4.1}$$

Therefore, for a high likelihood event (such as encountering rock in an excavation), low consequence event (the rock can readily be removed by blasting), the risk score is (4×2)=8 that is categorized as "medium risk" on Figure 4.5.

To account for the full range of probabilities and consequences, a matrix can be drawn up from which risk scores can be determined and the project risks ranked. Figure 4.5 shows the risk matrix with Likelihood scores from 1 to 5, and Consequence scores from 1 to 5, and Risk scores that are the product of the Likelihood and Consequence scores.

The risk scores can be categorized as High, Medium and Low as shown by the shading on the matrix. A feature of most risk matrices is that if the consequence is severe or catastrophic, which usually means that the hazard can result in injury or death, then the project is deemed to be high risk even if the probability of the event occurrence is very low or rare. For example, earthquakes, even in high seismic areas, are rare events but are also likely to be very destructive such that they are rated as high risk or very high risk events. Figure 4.6 shows a road in Turkey damaged by earthquake ground motions; earthquakes are a common cause of slope failures (Jibson, R.W., 2013).

Likelihood						
	5 High	5	10	15	20	25
	4	4	8	12	16	20
	3	3	6	9	12	15
	2	2	4	6	8	10
	1 Low	1	2	3	4	5
		1 Incidental	2	3	4	5 Catastrophic
		Consequence				

RISK
High
Medium
Low

Figure 4.5 Risk matrix showing scores for likelihood of occurrence and consequence of events, and corresponding risk scores.

Figure 4.6 Sinkhole generated in karstic terrain by magnitude 7.8 Turkey – Syria earthquake in February 2023 (image by T. K. Southam).

Mitigation – a final step in the qualitative risk analysis is to assess the overall project risk with respect to geotechnical issues. The purposes of assessing project risk may be to decide how much additional geotechnical investigation should be carried out, or what the preferred procurement method for the project is. For example, if the geotechnical risk is high, it may not be appropriate to have a Design-Build project because the contractors will need to build significant contingencies into their bids to account for unforeseen subsurface conditions.

The overall project risk can be assessed by first preparing a risk register of all the potential risks on the project, and then determining the risk score (risk = likelihood · consequence) for each item in the register. The total risk for the project can then be estimated by adding the score for each of the identified risks. This summary of the project risks will clearly show the anticipated highest risks on the project, and where further analysis and possible mitigation measures should be concentrated. If the total project risk is very high, it may be decided that the project should be abandoned, or significantly restructured with lower risk.

A further step in the qualitative risk assessment is to examine the mitigation measures that would be required to reduce the risk scores of the highest risk hazards to more acceptable residual levels. Mitigation measures can be expressed in terms of both possible cost and time to mitigate

Table 4.2 Example of project risk scores, and mitigation costs and duration

		Risk mitigation			
Risk	*Likelihood score (L)*	*Consequence score (C)*	*Risk score (L x C)*	*Cost to mitigate risk ($)*	*Time to mitigate risk (duration)*
Landslides	3	5	15	$300,000	4 months
Contaminated ground	4	4	16	$200,000	2 months
Rock and boulders	3	1	3	$50,000	1 month
Settlement	1	3	3	$0	
Liquefaction	1	2	2	$0	
Total project risk			39		
Total mitigation cost				$550,000	
Total mitigation time					7 months

the risk because the project team can probably estimate these values with adequate reliability for this preliminary stage of the project. Addition of the mitigation costs and times would then show the level of effort that would be required to manage the project risks, and where these costs could be prioritized to have the greatest consequence on successful project execution.

Table 4.2 shows a risk register of five risks, of which Landslides and Contaminated ground have the highest risk scores that are consistent with the high risk category of these two hazards as listed in Tables 3.3 and 3.4. The risk score for these two hazards of 15 and 16 can be referenced to the risk matrix shown in Figure 4.5 where they are rated as high risk, while the other three hazards are rated as low risk.

The estimated total mitigation cost for Landslides and Contaminated ground is $500,000. For landslides, mitigation may involve drainage, unloading the top of the slide or buttressing the toe that all reduce the likelihood of a slide, or moving structures below the slide to reduce the consequence of a slide. For contaminated ground, mitigation may involve more detailed investigations, removal and treatment, or building a treatment faculty for contaminated material that is discovered during construction.

It may also be possible to optimize mitigation measures. For example, for sites where drilling is planned to investigate contaminated soils, the investigation could be combined with drilling to investigate the rock conditions. This would also mean that the duration of the mitigation work could be reduced by a month.

4.5 QUANTITATIVE RISK ANALYSIS

As the project progresses from conceptual design to preliminary engineering and design development (Figure 4.3), the Qualitative Risk Analysis will usually be up-dated to a more precise Quantitative Risk Analysis. Updating the risk register for Quantitative Risk Analysis relies less on estimates and judgement and the Likert scores that are used early in the project, and more on numeric values for the probability of occurrence of events and monetary values for consequences of these events (Dai, F. C., Lee, C. F., & Ngai, Y. Y., 2002).

Quantitative risk analysis can be carried out deterministically or probabilistically, as described below.

4.5.1 Deterministic analysis

Deterministic analysis involves selecting single values for each of the parameters that define the identified risks. That is, the risk register would express the risk of occurrence as an annual or lifetime probability (P), the consequence of an event as a cost $\$_c$, and the mitigation as a cost $\$_m$, for which the total risk cost would be: $[(P \times \$_c) + \$_m]$.

For example, consider a proposed urban development site with an area of 10 km^2 in a region that has a history of landslides. For these site conditions where the main geotechnical risk for the development are landslides, determination of the landslide risk would require that numeric values be defined for the parameters discussed below. These values would be the average, or most likely values, estimated from available information.

a. **Probability of occurrence** – historical records of the area can be used to determine the annual probability of a slide, or the probability of a slide during the life of the project. If landslides are considered to be a site risk because the project is in a landslide-prone region, presumably records exist of slides that have occurred in the past. The hazard area is in a river valley where the geological materials are soft sediments and the lower part of the slopes are subject to river scour. Site investigations comprising geological, topographic and river hydrology studies show that the length and width of the valley that is at risk from landslides is about 40 km long and 8 km wide, with an area of about 320 km^2. If eight slides have been recorded in the last 100 years along the river valley, then the annual probability of a landslide in the valley is:'

$$P_{(annual)landslide} = 8/100 = 0.08 \tag{4.2}$$

If it is assumed, for simplicity, that that geological conditions with respect to landslide risk are uniform throughout the region, and that a landslide in the future could occur in an area of a past landslide, then any future landslides will occur at the same annual probability of 0.08 and could occur anywhere within 320 km² area of river valley.

If the area of the proposed development is 10 sq. km, then the annual probability that any future landslides could occur in the area of the development could be estimated as:

$$\begin{aligned} P_{(annual)development/landslide} &= \left[(10/320) \times (8/100)\right] \\ &= 0.0025, \ or\ 2.5E-3 \end{aligned} \tag{4.3}$$

A more comprehensive risk analysis would account for the condition that the valley slopes may become progressively more stable after each failure so that the annual probability of failure gradually decreases.

b. **Temporal probability** – the probability of residences being inhabited during a landslide will depend on their occupancy time, and the warning that may be available of hazardous slope movement occurring. That is, if the residences are occupied full-time and are not summer cabins, then the temporal probability of injury or loss of life when the landslide occurs is 1.0. However, if the landslide movement is slow and/or is monitored to provide a warning of deteriorating stability, then the residents can be evacuated before their properties are destroyed so no injury or loss of life occurs, and the temporal probability is close to zero. That is, it will be assumed that time will be available to evacuate the residents before hazardous conditions develop. This contrasts with debris flows that may occur with little warning, and move at high velocity so that residents on the runout fan could have insufficient time to evacuate.
c. **Consequence of a slide** – for this urban location, it may be possible to determine the consequence of a landslide in terms of the number of houses that would be destroyed or damaged, and apply a cost of replacement or repair. Based on the number of buildings within the 10 km² landslide hazard zone, it can be estimated that the cost of damage in the event of ground movement is $3,000,000.
d. **Probability consequence** – the product of the probability of occurrence and the consequence of the slide is the Expected Value (EV). For example, if the annual probability of a landslide in the development area is 0.0025, and the consequence of the slide is damage of $3,000,000, then the annual Expected Value of damage is: [(EV_{damage} = 0.0025 × $3,000,000) = $7,500 per year].
e. **Risk acceptance** – the calculated risk of landslide occurrence can be compared to acceptable societal risk, or perhaps risk deemed acceptable by local planning authorities. An example of a risk matrix that can be used to evaluate risk is shown in Figure 1.3 – if the annual frequency of a landslide in the development area is 2.5E-3, and even if no fatalities occur, then the risk for the development is deemed to be

"unacceptable". For these conditions of unacceptable risk, mitigation measures will be required if the development is to be approved, as described below.

f. **Risk mitigation** – the cost, time and likely success of mitigation of the landslide risk would be evaluated because, having identified an unacceptable risk, it would be the responsibility of the risk management team to assess how the consequence of the landslide could be mitigated. If the potential landslide has an area of 10 km^2, the remedial costs are likely to be substantial. For example, the consequence could be eliminated if development were prohibited and the government buys out the development at well-defined market prices (Just Case: Hazard Development at Garibaldi, 1989). If the site has been developed when the landslide potential is recognized, then possible stabilization measures may involve drainage, that will require on-going maintenance, or unloading the crest of the slide; both mitigation measures are usually costly endeavours for a large slide.
g. **Residual risk** – when mitigation has been carried out, the initial risk is reduced to a residual risk that is calculated from the new probability of occurrence and the new consequence. If the mitigation has been to unload the upper part of the slide at a cost of \$800,000, and this is considered to have reduced the probability of failure by an order of magnitude to 0.00025 (2.5E-4), but the consequent cost of \$3,000,000 is unchanged because the residences have not been moved, then the annual residual Expected Value is: [(EV_{damage} = 0.00025 × \$3,000,000) = \$750 per year].

   The reduction in the annual EV from \$7,500 to \$750 over the life of the development would need to be balanced against the onetime remediation cost of \$800,000.

   A quantitative analysis of these options to determine the optimum course of action could be carried out using decision analysis that takes into consideration the uncertainties in both the landslide risk and the mitigation costs. Decision analysis is discussed in Chapter 7.
h. **Monitoring and retiring risk** – if the development proceeds, risks would be monitored and modification made to the landslide probability and consequences, as deemed appropriate. For example, it may be found that the materials in the potential slide area have higher shear strength than anticipated, in which case the extent of the unloading work could be reduced. Finally, at the completion of the mitigation work, it may be decided that the residual risk of a landslide is no longer significant and that the risk can be retired.

The eight steps in the quantitative risk management of the landslide described above provide a useful analysis of the value of mitigating the landslide risk using readily defined average values of probability, and of consequence and mitigation costs. This deterministic analysis can be a good starting point for quantitative analysis and a guideline on whether more detailed probabilistic analysis is required.

4.5.2 Probabilistic analysis

Investigation of geotechnical site conditions will usually show uncertainty in values of the parameters that will be used in design. Furthermore, the level of uncertainty will often differ for different parameters – for example, with shear strength, cohesion that is often more difficult to define than friction angle, will have a higher degree of uncertainty than friction angle. If it is considered that selected single, average parameter values used for deterministic analysis are too uncertain for detailed design, then probabilistic analysis can be used in which uncertainty in the values can be better quantified, as described in this section.

The basis of probabilistic analysis is to define all parameters for which uncertainty exists, as probability distributions rather than single, average values. Quantification of uncertainty is discussed in Section 1.6 where Figure 1.7 shows four commonly used probability distributions that may be suitable to define the possible ranges of parameter values. These four distributions as well as the uniform distribution, and the conditions for which they may be used, are as follows:

a. **Triangular** – distribution defined simply by maximum, minimum and most likely values that can be useful when field data is limited, or absent, in which past experience and judgement can be used to define the three parameters. The distribution can be symmetrical, or asymmetric to account for bias towards the maximum or minimum values.
b. **Normal** – this is the most commonly used distribution and is defined by the average value and the standard deviation (see equations 1.2 and 1.3) – it is a symmetric distribution that extends to infinity in both directions, and includes negative values.
c. **Beta** – distribution where the maximum and minimum values, and the skew (or degree of asymmetry) are defined. This distribution is useful for geotechnical parameters where the maximum and minimum values can be reasonably well defined, and it is expected that the most likely value is closer to one or the other extreme. It is considered that the Beta distribution is particularly useful in avoiding unrealistically high or low values, such as rock mass friction angles of 60° or 10° respectively, that are generated by normal distributions.
d. **Log-normal** – distribution for which negative values cannot occur, but the maximum value is infinite, and asymmetry is possible. This

distribution is useful for defining such geological features as the persistence and spacing of joints in a rock mass, and the dimensions of rock falls. For these parameters, the average value may be a relatively low value, such as a joint persistence of 2.5 m, but a persistence of 10 m is possible although it has a low frequency of occurrence. Quantification of rarely occurring, high persistence joints may be important because this would define the possibility of occasional, large scale slope failures.

e. **Uniform** – distribution with defined maximum and minimum values, and a constant probability of occurrence between these two values. An example of an application of a uniform distribution is the study of rock fall hazards where a rock may fall with equal probability from any height on the slope if the geological and topographic conditions are similar at all locations on the slope.

Calculations carried out with parameters that are defined by probability distributions, rather than single values, can be readily performed using Monte Carlo analysis as described below.

4.5.3 Monte Carlo analysis

Once the probability of occurrence and consequence parameters have been defined by appropriate probability distributions, it is possible to multiply the two distributions to calculate a distribution for the project risk. Because it is not readily possible to mathematically combine different types of distributions, these calculations are most conveniently carried out by Monte Carlo analysis using programs such as @Risk™ or Crystal Ball™. Monte Carlo analysis, which is described in Section 2.3.4, involves using random numbers to generate, from the probability distributions for the occurrence and consequence of an event, a value for each of the two parameters and multiply the two values to calculate the risk. By running these analyses thousands of times, each time with different occurrence and consequence values generated by the random numbers, a probability distribution is generated for the calculated risk.

An example of Monte Carlo analysis to calculate probabilistic landslide risk is discussed below; the analysis follows the deterministic scenario described in Section 4.5.1. Figure 4.7 shows estimated probability distributions for the area of the landslide hazards, the historic number of slides and the cost of damage due to future landslides. Uncertainty in the values of the risk parameters and their probability distributions are as follows (see Figure 4.7):

a. **Number of recorded slides** = 8. Landslide records over 100 years may not be reliable because, for example, records could have been lost, or slides may not have been recognized or recorded. It could be estimated

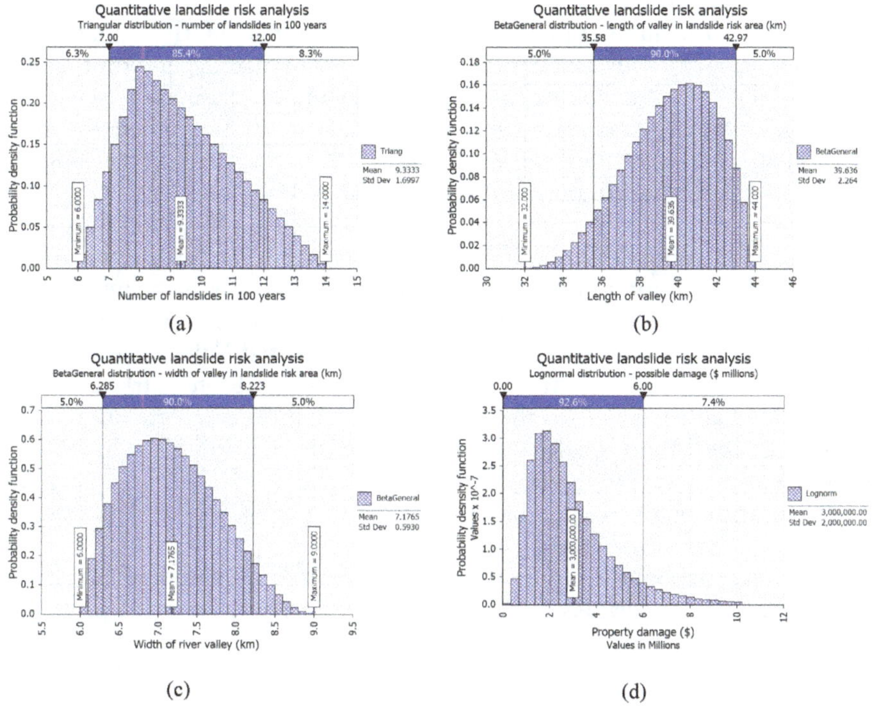

Figure 4.7 Probability distributions for landslide risk analysis, input values: (a) number of historic slides; (b) and (c) length and width of valley, landslide hazard area; (d) cost of damage due to landslide.

that the most likely number of slides is eight, but that as many as 14 and as few as six may have occurred. This uncertainty could be expressed as a triangular distribution, as shown in Figure 4.7a).

b. **Area where landslides have been recorded** = 320 km^2, represented by a 40 km length of the river valley with an average width of 4 km on each side of the valley. However, locations of the early landslides could be uncertain because they cannot be clearly identified in the field. Study of aerial photographs, geology and topography indicates that it is possible that past landslide activity could be in the range of 32 to 44 km length of the valley, and 6 to 9 km for the width of the valley. These uncertainties could be expressed as Beta probability distributions with maximum and minimum values as shown in Figure 4.7b and c, with asymmetrical, skewed distributions indicating dimensions that are considered to more likely than the average dimension.
c. **Years of landslide history** = 100. It is assumed that only 100 years of records are available.
d. **Area of proposed development** = 10 sq. km. This is a fixed area based on the development permit.

e. **Damage caused by landslide=$3,000,000.** This value is uncertain because the number of houses that may be damaged, the extent of the damage, can only be estimated. Damage will also depend on the vulnerability of the houses to slope movement. Uncertainty in the damage cost can be expressed as a log-normal distribution where the most likely cost is $2,000,000 but there is a 6% chance that the damage could be greater than $6,000,000 (see Figure 4.7d)).

The input probability distributions shown in Figure 4.7 represent the uncertainty in the information on landslide conditions. These distributions can be used to compute distributions for the annual probability of a landslide in the development area, as well as the annual Expected Value (EV) of damage to buildings due to a landslide (see Figure 4.8). Monte Carlo analysis, as described above, has used the input distributions to calculate output distributions; the Monte Carlo analysis and the distribution plots have been generated by the software @Risk™

A discussion on the output distributions shown in Figure 4.8 is as follows:

a. **Annual probability of a landslide in the valley** – based on historical records of landslides collected over 100 years, the annual probability of a landslide in the valley is the number of landslides divided by

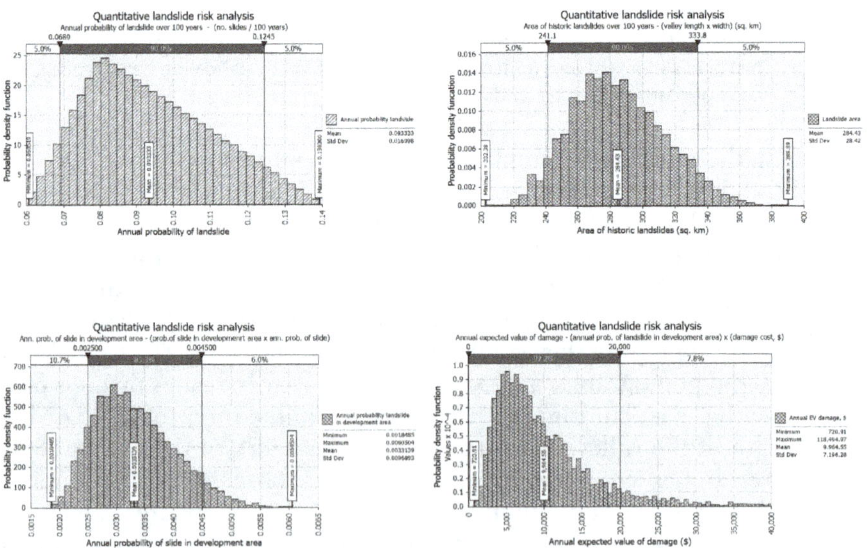

Figure 4.8 Landslide risk analysis, output distributions: (a) annual probability of landslide; (b) probability distribution of landslide area; (c) annual probability of slide in development area; (d) annual Expected Value (EV) of damage due to landslide.

100 years; Figure 4.8a shows that this a triangular distribution defined by the triangular estimate for the number of historic landslides.

$$P_{landslide} = \frac{Number\ of\ slides\ in\ 100\ years\ of\ records}{100\ years} \tag{4.4}$$

The mean annual probability of a slide in the valley is: $P_{\text{valley}}=0.09$ slides per year, with maximum and minimum values of 0.14 and 0.06 respectively.

b. **Historic landslide area** – the probability distribution for the possible landslide area is generated by multiplying the two Beta distributions for the length and width of landslide hazard area along the valley (see Figure 4.7b and c). The calculated distribution for the area shows that the mean, maximum and minimum values for the landslide area are 283, 383 and 213 km^2 respectively, and that 90% of the area is between 240 and 334 km^2 (Figure 4.8b).

c. **Annual probability of slide in development area** – assuming that the whole area of the valley has the same landslide risk, then the probability of a slide in the 10 km^2 development area is:

$$P_{slide\ in\ development\ area} = \frac{areas\ of\ developemt}{area\ of\ valley\ hazard\ area} \tag{4.5}$$

The annual probability of a slide in the development area is the product of the annual probability of a slide in the valley (P_{valley}), and the probability that the slide will be in the development area ($P_{\text{development area}}$). The resultant is shown in Figure 4.8c where the mean annual probability is 0.003, maximum and minimum annual probabilities are 0.006 and 0.002 respectively.

d. **Annual Expected Value (EV) of landslide damage** – the product of the distribution of the possible damage costs (Figure 4.7d), and the annual probability of a slide in the development area (Figure 4.8c) is the annual EV of damage resulting from landslides (Figure 4.8d). Figure 4.8d shows that the mean annual expected value is $10,000. The annual minimum expected damage value is only $720, and the annual maximum value is $118,000. Because the damage expected values are based on a lognormal distribution where the maximum cost is infinity, the calculated maximum expected value at $118,000 is high, but its probability of occurrence is very low. The distribution in Figure 4.8d also shows that the probability is 92% that the annual expected value is less than $20,000.

The calculated annual expected value of damage can be used to assist in the decision to proceed with the development, and if so, how much should be spent on mitigation measures. Selection the optimum decision for the site can be achieved by the use of Event trees (Chapter 6) and Decision Analysis (Chapter 7) where the cost of upfront remediation work, such as a drainage system can be compared with reduction in the annual probability of a landslide occurring in the future, and the reduced expected value of damage to buildings by landslides.

4.6 BAYESIAN ANALYSIS

Section 4.3 above discusses how risk studies progress from qualitative evaluation in the early stages of the project when little information is available, to quantitative analysis using numerical values developed when the investigation program provides site-specific data. At all stages of this work, uncertainty in geotechnical parameters will exist that can be addressed in design using factors of safety, or conservative assumptions. However, uncertainties can also be represented by probability distributions that help to quantify uncertainty and can be incorporated into probabilistic analysis.

4.6.1 Principles of Bayesian Analysis

The progress of project development usually involves first using existing data (*Prior model*), and then collecting additional data as required for final design (*Posterior model*). This sequence means that it is beneficial for design to have a rigorous means of updating the probability distributions, using new information as it becomes available and is incorporated into the site data. Incorporation of new data generally results in a progressive decrease in the level of uncertainty, provided that the new data has consistency with the original data. A suitable method of explicitly updating probabilities using site data, as well as theoretical models and expert opinion, is to use Bayesian analysis as described below.

A general description of Bayesian analysis is that it *finds the probability of causes by examining the effects*. For example, in slope stability analysis the effect could be slope instability, and examination of slope conditions may find the cause of instability to be excessive water pressure. If the magnitude of the water pressure is uncertain, the pressure would be expressed as a probability distribution to calculate the probability of water pressure causing the slope to fail.

Bayesian analysis involves the following six steps (Straub, D., & Papaioannou, I., 2017; Baecher, G. B. & Christian, J. T., 2003):

a. Establish an initial prior model using existing data (*prior distribution*).
b. Compute the reliability and risk based on the Prior model.
c. Describe new observations and data that are also defined by probability distributions (*likelihood*) – these data make up the hypotheses that characterize the actual site conditions.

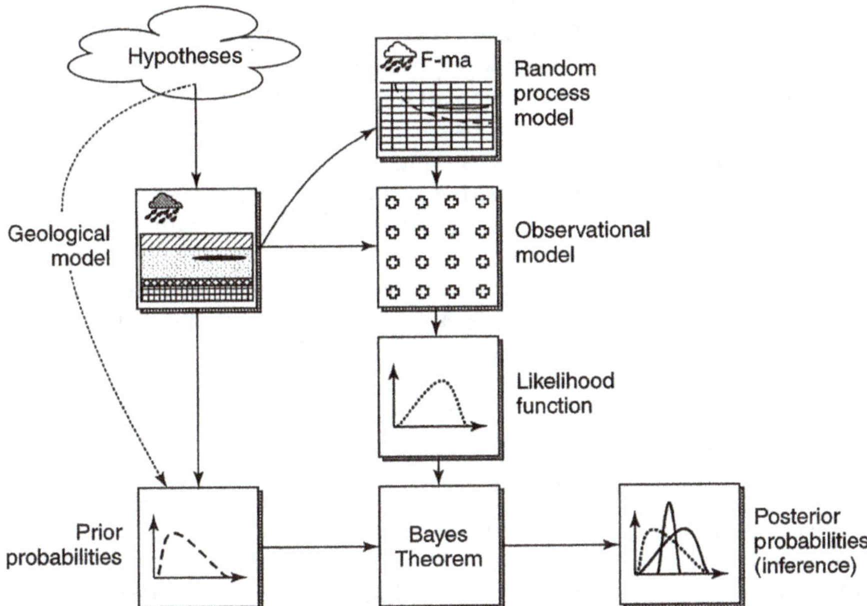

Figure 4.9 Illustration of the use of Bayes theorem to analyze site data hypotheses and develop Posterior probabilities (Baecher, G. B. & Christian, J. T., 2003).

d. Update the prior model (*posterior distribution*).
e. Update the reliability and risk.
f. Communicate the result.

This sequence of activities is shown graphically in Figure 4.9 where the concept of probability and likelihood is illustrated in the context of Bayesian analysis. Existing information on the site and geological model defines the Prior probabilities. Collection of new data comprising the geological model, estimates of site conditions (random processes) and observations of site conditions form the hypotheses characterizing site conditions; analysis of these site conditions defines the Likelihood function. Following this, Bayes theorem can be used to combine the Prior Probability and the Likelihood function to calculate the Posterior probability.

Note that the Likelihood function as used in Bayesian analysis discussed in this section differs from the term "likelihood" that is used in the risk management matrices shown in Figures 4.2 and 4.5. In the risk matrices, the term likelihood refers to the possibility of an event or activity occurring, and is used to indicate that occurrence of the event is a qualitative estimate rather than a quantitative probability number developed using site-specific data.

The distinction between probability and the Likelihood function is fundamentally important to Bayesian analysis and to statistical inference. That is, probability attaches to possible results and the probability of their occurrence, while the Likelihood function attaches to hypotheses and is the probability of a measurement outcome (Fisher, R. A., 1921).

In geotechnical engineering, the primary use of Bayesian analysis is to develop sets of design parameters that incorporate uncertainty in their values that can be up-dated as more information is collected and the project develops.

Information collected in investigation programs can be classified as events, with measurements, such as the depth of the water table denoted by the symbol *W*. The geotechnical model for the stability of slopes, defined by the factor of safety, *FS* is given by the ratio:

$$FS = \frac{(capacity,\ resisting\ forces)}{(demand,\ displacing\ forces)} \tag{4.6}$$

All events can be expressed by the terms ***FS*** that are a function of ***X***. For example, the event *FS*="loss of stability" corresponds to the condition $[FS<1.0]$ such that the factor of safety is less than 1.0. The goal of Bayesian analysis is to quantify the consequence of the measurement outcome *W* on the design parameters "|", and ultimately on the event of interest, *FS*.

In Bayesian analysis, quantification of the consequence of design parameters on stability is carried out by computing conditional probability as follows. For event ***FS***, slope instability, the conditional probability of slope instability is $P(FS|W)$, given that the information *W* on the groundwater condition has been collected; the symbol "|" represents the term "given that".

From the site investigation program, the new information *W* may show that the level of the water table is less than that predicted for the Prior distribution ($P(FS)$ so that the new distribution will be less than the initial estimated probability of *FS*, that is, $P(FS|W)<P(FS)$.

The conditional probability of slope failure *FS*, given the measurement of the water table $P(W)$, is defined as:

$$\mathrm{P}(FS|W) = \frac{P(FS \cap W)}{P(W)} \tag{4.7}$$

where the intersection symbol "∩" represents the condition that two independent events *FS* and *W* occur together. This condition can be illustrated by a Venn diagram of two overlapping conditions – the slope failure and the water table level where the probability of the water table being high enough to cause slope instability is proportional to the area of the overlap. If no water pressure condition exists that can cause slope instability, then the distributions of slope instability and water pressure do not overlap.

Following Bayes theorem, the conditional probability of *FS* (e.g., slope instability) given that *W* (e.g., groundwater level) has been measured, is defined by the posterior probability:

$$\mathrm{P}(FS|W) = \frac{(P(W|FS) \cdot \mathrm{P}(FS))}{\mathrm{P}(W)} \tag{4.8}$$

where the terms have the following meanings:

- $P(FS)$ Prior probability of slope instability
- $P(FS|W)$ Posterior probability of instability, given that water table has been measured
- $P(W|FS)$ *Likelihood* of making observation *W* given stability conditions *FS* (see Section 4.6.3 below)
- $P(W)$ probability distribution of the observations of the water table.

A component of the posterior probability calculation is the normalizing constant *a* which ensures that the posterior distribution integrates to 1.0, and is required so that the total area under a probability distribution curve represents all possible values of the parameter.

Bayes rule defining the Posterior probability and incorporating the normalizing constant *a* is as follows:

(Posterior probability, $P(FS|W)) = (a) \cdot$ (Likelihood function, $P(W|FS)) \cdot$ (Prior probability, $P(FS)$) (4.9)

Combining equations (4.8) and (4.9) shows that:

$$a = \frac{1}{P(W)} \tag{4.10}$$

That is, constant *a* is an indication of how likely the observation of the water table level is to occur.

4.6.2 Bayesian analysis for geotechnical engineering

Application of Bayes theorem to geotechnical investigations is illustrated as follows.

A geotechnical design for stabilization of a slope requires values for the shear strength of the material in terms of the friction angle, assuming that the cohesion is zero and water is not present. At the start of the project, estimates are made for the values of the friction angle based on experience and information from nearby sites. Because of uncertainty in site conditions, the value of the friction angle has a range represented by a probability distribution. This distribution for the friction angle of the random variables can be expressed as a function **X** and is termed the **Prior Probability.**

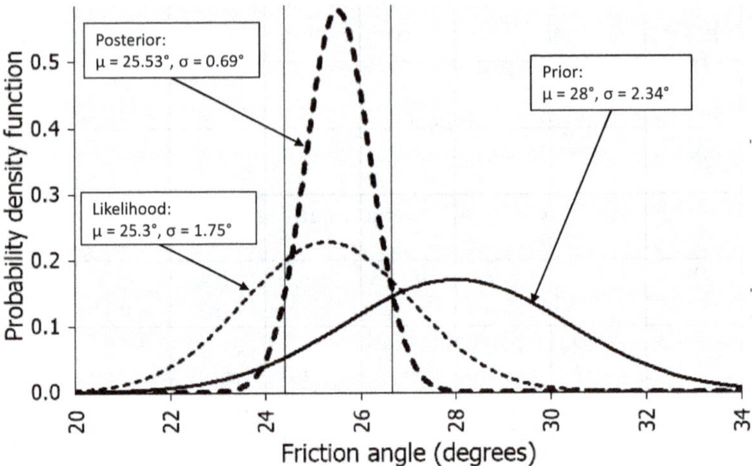

Figure 4.10 Probability distributions for Prior, Likelihood and Posterior information illustrating Bayes theorem for the mean friction angle ($\phi°$) (Straub, D., & Papaioannou, I., 2017).

The next step in the project is to carry out a site investigation to obtain site-specific data that involves drilling to obtain samples and laboratory testing of the friction angle. However, because of site access restrictions, the investigation program is limited and uncertainty in site conditions remains after the investigation work. For this situation, Bayes theorem can be applied to combine the initial estimate of the design parameters (**Prior Probability**) with the data obtained in the site investigation (**Likelihood**) to prepare an updated data set (**Posterior Probability**) (Figure 4.10).

The following comments apply to the three distributions shown in Figure 4.10, where the three sets of data applicable to the site are defined by normal distributions.

- **Prior distribution,** $f'_{\mu\phi}(\mu_\phi)$ – before the site investigation is carried out, available data from the area shows that mean and standard deviation of the soil friction angle are:

  Prior normal distribution: $\mu_\phi=28°$ and $\sigma_\phi=2.34°$

$$f'_{\mu\phi}\left(\mu_\phi\right)=\frac{1}{2.34\cdot\sqrt{2\pi}}exp\left[-\frac{1}{2}\left(\frac{x-28}{2.34}\right)^2\right] \tag{4.11}$$

- **Likelihood function,** $L(\mu_\phi)$ – a site-specific investigation shows that six samples ($n=6$) have values for the friction angle of ϕ_i for which the normal distribution is defined by:

For $\phi_{i=1}$ to $\phi_{i=6}$: $\mu_\phi=25.3°$ and $\sigma_\phi=1.75°$.

The Likelihood function for these six data points is calculated from:

$$L(\mu_\phi) = \frac{1}{\left(\sigma_\phi \cdot \sqrt{2 \cdot \pi}\right)^n} exp\left[-\frac{1}{2}\sum_{n=1}^{n=6}\left(\frac{\phi_n - \mu_\phi}{\sigma_\phi}\right)^2\right] \tag{4.12}$$

$$L(\mu_\phi) = \frac{1}{\left(1.75° \cdot \sqrt{2 \cdot \pi}\right)^6} exp\left[-\frac{1}{2}\sum_{n=1}^{n=6}\left(\frac{\phi_n - 25.3°}{1.75°}\right)^2\right]$$

- **Posterior** $f''\mu\phi(\mu_\phi)$ – the Posterior distribution is calculated from the Prior distribution, the Likelihood function and the normalizing constant a (see equation 4.10) as follows:

$$f''_{\mu\phi}(\mu_\phi) = a \cdot L(\mu_\phi) \cdot f'_{\mu\phi}(\mu_\phi) \tag{4.13}$$

Figure 4.10 illustrates the three probability distributions showing how the distribution for the Posterior probability is higher and narrower than that for the Prior probability indicating the reduced uncertainty after the data collection program.

4.6.3 Likelihood function

Bayes rule defines how a prior probability $P(FS)$ is updated to a posterior probability $P(FS|W)$ when making the observation W (equation 4.8). The observation is described by the Likelihood term $P(W|FS)$ that is defined as the conditional probability of making measurements W given a particular system state $P(W|\textit{system state})$.

The Likelihood function for a set of data such as the friction angle ϕ can be calculated using equation (4.12) that is a function of the number of samples, n in the data set. For a data set comprising 20 measurements of the friction angle ranging from 22° to 30° that have mean and standard deviation values of: μ_ϕ=25.36° and σ_ϕ=2.03° respectively, Figure 4.11 shows plots of the Likelihood function for the number of samples: n=1, 3, 5, 10, 20. Figure 4.11 shows how increasing the number of samples decreases the width of the Likelihood function, indicative of the reduced uncertainty in the design data, for this example where the 20 samples are in a narrow range of ±3°.

4.6.4 Bayesian updating calculations

Bayesian updating is computationally demanding and, in general, is performed numerically which requires that efficient algorithms be used. The

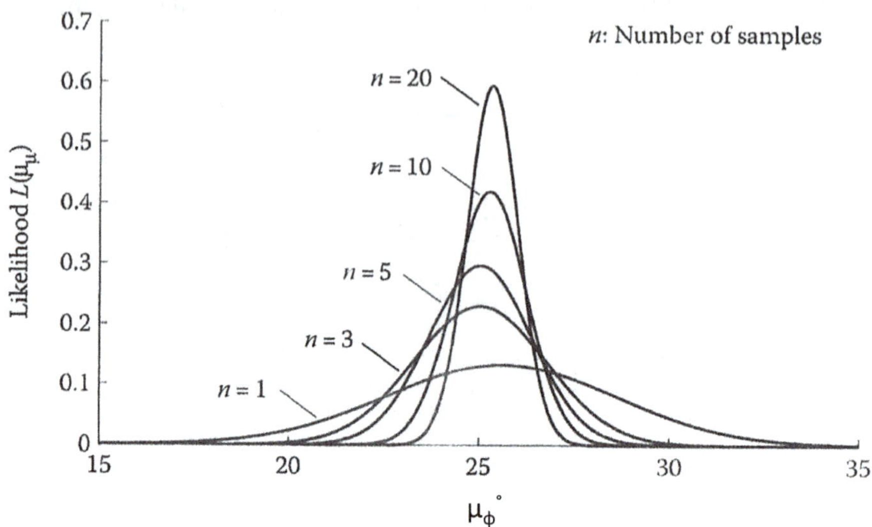

Figure 4.11 Likelihood function (equation 4.12) of mean friction angle, μ_ϕ for different number of samples, n (Straub, D., & Papaioannou, I., 2017).

following are brief descriptions of two computation methods applicable to Bayesian updating; these methods are described in more detail by Straub and Papaioannou (2017) and the cited references.

a. Markov chain Monte Carlo (MCMC) – MCMC is a powerful approach for generating samples from distributions from which direct sampling is difficult. The main advantage of MCMC methods is that they do not require complete specification of the distribution that is being sampled. This is particularly useful for Bayesian updating whereby the posterior distribution is known only up to a normalizing constant. The basic idea of MCMC is to construct a stationary Markov Chain with invariant distribution equal to the target distribution.
   Details of Markov Chain analysis methods are discussed by (Gelman, A., 2004) and (Tierny, L., 1994)
b. Sequential Monte Carlo – the basic idea behind this method is to gradually translate samples from the prior distribution to samples from the posterior distribution through a sequential reweighting operation. Reweighting is based on importance sampling from a sequence of distributions constructed such that they gradually approach the target posterior distribution (Del Moral, P., Doucet, A., & Jasra, A., 2006; Chopin, N., 2002; Neal, R. M., 2001).

4.6.5 Bayesian updating software

To facilitate Bayesian updating calculations, software is available that incorporates specific Bayesian functions, an example of which is STATA (Stata Corp LLC, 2021). An advantage of this software is that the probability distributions are not limited to normal distributions so that distributions can be selected that best suit the project data.

4.7 ANALYSIS OF CONSEQUENCES

The risk matrices in Figures 4.2 and 4.5 show the likelihood of an event occurrence in terms of the annual probability, or activity likelihood ranging from "rare" to "almost certain". The corresponding consequences of these events or activities occurring are also shown in the matrices. Detail that is provided on the consequences will depend on the size of the project and the possible severity of adverse outcomes. For small projects, consequence could be ranked in terms of the five point Likert scale ranging from minimal/very mild (1 point) to extreme/very critical (5 points), as shown in Table 4.1 and Figure 4.5.

On large and complex projects, a more detailed and broader range of consequences, or negative outcomes, may be considered consistent with project conditions (Figure 4.2). Examples of types of consequences, that extend from "incidental" to "catastrophic", are:

- **Health and Safety** – consequences ranging from slight consequence that is recoverable within days, to accidents resulting in multiple fatalities. Management of these consequences usually involves use of wearing mandated safety clothing (e.g., hard hat, Hi-Viz vests), safety training courses at the start and during the project, safety audits of work practices including reporting of incidents, and presence on site of first aid personnel and equipment.
- **Environment** – consequences ranging from localized short term impact, recoverable within days, to irreplaceable loss of a species. Management of these consequences usually involves taking inventories of site conditions prior to construction (such as wildlife and vegetation), mandating construction methods that are environmentally benign (such as use of non-toxic hydraulic fluids), and having rapid response capabilities for a spill clean-up.
- **Social and Cultural** – consequences ranging from negligible consequence to complete loss of social and cultural values. Management of these consequences usually involves study of local social and cultural

values with relevant stakeholders prior to construction, design of the project to avoid or mitigate consequence to these values, and close monitoring during construction of these values to minimize the risk of disturbance and destruction.

- **Economic** – consequences ranging from brief work stoppages, to business interruption losses exceeding $10 million. Economic consequences can affect both external businesses such as disturbance to buried utilities and nearby buildings that are vulnerable to damage, and internal consequences such as schedule delays and cost overruns. Management of these consequences usually involves identification of vulnerable facilities prior to construction, preparation of designs to minimize consequences, and use of construction methods that limit effects outside the construction area, such as noise control.

  For industries such as electrical power generation facilities or public transportation systems that operate under strict regulations, two possible negative consequences of an event such as a landslide are listed at the end of Table 4.3. Management of these risks will probably require involvement of full-time risk management personnel who are familiar with the regulations and monitor operations to ensure compliance.
- **Regulatory** – consequences ranging from an event for which no regulatory involvement is required, to an event that results in a regulator imposed shutdown, a fine and substantial damage to the corporate image.
- **Government regulations** – consequences ranging from requirement to communicate/notify external agencies with no cost consequence, to company losing influence with external agencies resulting in need for costly changes to procedures.

Table 4.3 lists the six classes of consequences from Health and Safety to Regulatory Compliance, for six classes of severity ranging from incidental to catastrophic. The descriptions of consequences that are shown on this table would be customized for the type of event that is expected to occur. Events may be categorized as either natural events such as earthquakes or landslides, or operational activities such as structural instability or an accident. The descriptions of consequences should also be customized for the type of organization such as a government agency, or a private company depending, on the exposure of their operations to the public.

Table 4.3 Descriptions of possible consequences, negative outcomes (see also Figure 4.2)

	1	*2*	*3*	*4*	*5*	*6*
Indices	*Incidental*	*Minor*	*Moderate*	*Major*	*Severe*	*Catastrophic*
Health & Safety	No consequence	Slight consequence: recoverable within days	Minor injury or personal hardship; recoverable within days or weeks	Serious injury or personal hardship; recoverable within weeks or months	Fatality or serious personal long-term hardship	Multiple fatalities
Environment	Insignificant	Localized short-term consequence: recovery within days or weeks	Localized long-term consequence; recoverable within weeks or months	Widespread long-term consequence; recoverable within months or years	Widespread consequence; not recoverable within the lifetime of the project	Irreplaceable loss of a species
Social & Cultural	Negligible consequence	Slight consequence to Social and Cultural Values: recoverable within days or weeks	Moderate consequence to social & cultural values; recoverable within weeks or months	Significant consequence to social & cultural values; recoverable within months or years	Partial loss of social & cultural values; not recoverable within the lifetime of the project	Complete loss of social & cultural values
Economic	Negligible: No business interruption	<$10,000 business interruption loss or damage to public or private property	<$100,000 business interruption loss or damage to public or private property	<$1M business interruption loss or damage to public or private property	<$10M business interruption loss or damage to public or private property	>$10M business interruption loss or damage to public or private property

(*Continued*)

Table 4.3 (Continued) Descriptions of possible consequences, negative outcomes (see also Figure 4.2)

Social (government relations)	Event causes no external relations issues	Event leads to communication or notification to external agencies; minimal follow-up required and minimal or no cost consequence	Event leads to communication with external agencies with follow-up discussions and/or reporting required. May have cost consequence on future decisions.	Trust with external agencies eroded to point that cannot be rectified by enhanced communications. Will have a cost consequence on future decisions.	Loss of influence with external agencies that will have significant cost consequence on future decisions	Public enquiry held into event, with intense scrutiny by government agencies of the event, and of the organization's operations.
Regulatory (legal compliance)	Event occurs for which there is no regulatory involvement, and no corporate image consequence	Event occurs that results in the regulator requiring an informal meeting	Event occurs that results in the regulator issuing an order to comply. Moderate consequence to corporate image	Event occurs that requires multiple internal meetings and allocation of resources to revise practice or procedure to meet regulator's order. Significant negative consequence to corporate image	Event occurs that results in a regulator imposed shutdown of part or whole of system, substantial fines and very significant damage to corporate image.	Event occurs that results in very considerable penalties and prosecutions. International negative public attention to event.

4.8 CASE STUDIES – RISK HAZARDS AND CONSEQUENCES

As a follow-up to risk analysis discussed in this chapter, these issues are illustrated below in the four case studies.

4.8.1 Debris flow containment dam – Event risk

Risk management of debris flows involves study of the probability of occurrence in terms of the event magnitude-frequency relationship for the site, and the consequences in terms damage to downstream facilities and injury/death to persons (Skermer, N. A., 1984; VanDine, D. F. & Lister, D. R., 1983).

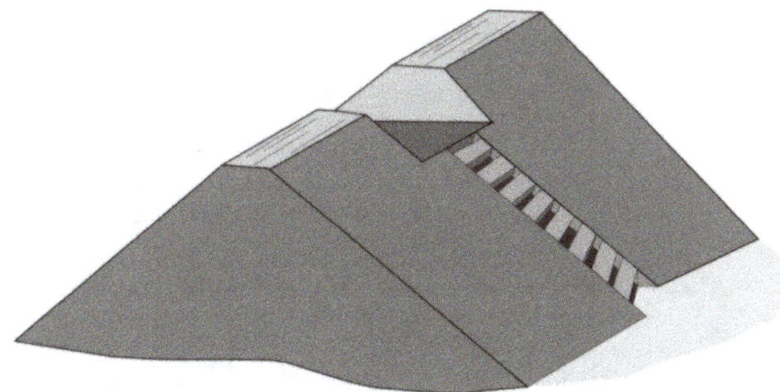

- **Hazard** – debris flows are a risk that are usually readily identified – a steep gradient stream with an accumulation of debris - sand to boulder size particles - in the creek bed, located in an incised valley that is subject to periods of sudden, high volume flows. Furthermore, debris from previous events accumulate as fans where the valley widens and the gradient becomes less steep. Debris flows are a highly fluid mixture of water, solid particles and organic matter that has the consistency of wet concrete and consists of 70% to 80% water. The solid materials range from clay and silt sizes up to boulders up to several metres in diameter, with organic matter comprising bark mulch and large trees. Flow velocities can be in the range of 3 to 5 m/s, with pulses as high as 30 m/s.
- **Consequence** – if a dam such as that shown in the sketch has been constructed to contain debris flows, a possible consequence for downstream facilities could be overtopping if the capacity of the dam is exceeded by the flow volume, or if an accumulation of previous small volume events has not been removed. Presumably, the risk of failure of the entire dam, due to erosion of the foundation or abutments for

example will be lower than overtopping. Despite the construction of a dam, a clear channel for creek flow should be maintained through the dam, and downstream of the dam.

4.8.2 Rock fall hazard – event risk

Construction of highways and railways in steep, mountainous terrain often requires excavation of rock cuts that need to be steep (often 76° or 0.25 H:1 V), and have narrow ditches with restricted catchment capacity, in order to limit excavation volumes. These conditions can result in rock fall accidents.

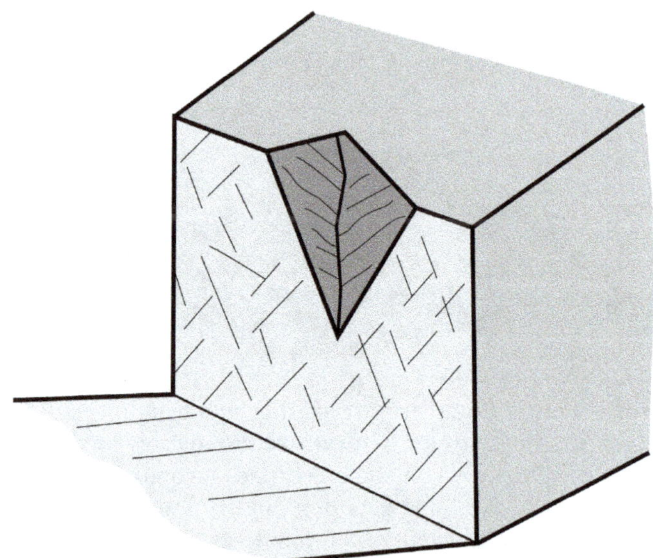

- **Hazard** – rock falls from the face of the cut, or soil slides from above the crest, may exceed the capacity of the ditch and spill on to the road or railway. Because sight distances are often limited by the curved alignment, it is unlikely that vehicles will be able stop or avoid impact with the fall. Vehicles are very vulnerable to rock fall accidents because even small falls can result in loss of control and the possibility of collisions with other vehicles. In contrast, locomotives are heavy, and are difficult to derail and dislodge from the track.
- **Consequence** – possible consequences of rock fall accidents are both direct and indirect. Direct consequences are damage to vehicles and

injury to occupants, while indirect consequences are delays to traffic, emergency response actions and legal costs in the event of claims for damages.

4.8.3 Tunnel stability – activity risk

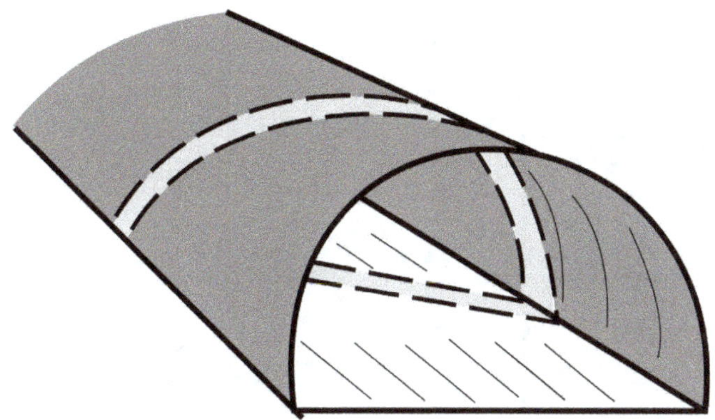

Tunnels that intersect faults can be a hazard, particularly if their presence is unexpected and their properties are unknown. The location and properties of a fault may be detected by surface drilling, but this could be difficult if the tunnel is deep and no surface expressions of the geology at the elevation of the tunnel are visible. An effective means of managing risk of fault intersection is to drill probe holes ahead of the face, although this will slow the advance rate.

- **Hazard** – possible hazards when a tunnel intersects a fault are rock falls, high inflows of groundwater and possibly tunnel collapse, all of which can be a hazard to mining personnel, particularly because these events could occur with little warning.
- **Consequence** – in addition to consequences of injury to personnel and damage to equipment, it is likely that costly remedial work will be required such as extensive support of the poor ground, grouting for groundwater control and further probe drilling ahead of the face. These activities will take time and slow the tunnel advance rate, and add to project cost.

4.8.4 Dam foundation- activity risk

Installation of high capacity, tensioned rock anchors in concrete gravity dams can be required to improve the seismic stability of the structure. The project involves drilling carefully aligned, steeply inclined holes through the dam and into the foundation rock, following which multi-strand or rigid bar anchors are installed, grouted in the foundation and then tension is applied at the head of the anchor on the top of the dam to increase the normal stress between the base of the dam and the foundation.

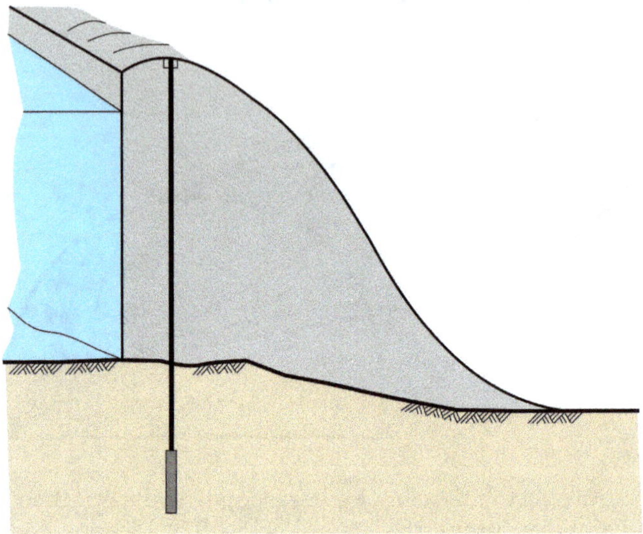

- **Hazard** – it is unlikely that the anchoring operations will be a hazard to personnel. However, possible construction hazards are that it is not possible to maintain alignment of the drill holes, that the anchoring grout leaks into the rock surrounding the bond zone, or that the bond zone slips during tensioning.
- **Consequence** – possible consequences of the hazards discussed in the previous paragraph are construction delays and cost overruns. Remedial work may be to modify construction methods and/or revise the design.

Chapter 5

Risk evaluation and acceptance

Chapter 4 discusses methods for identifying the hazards and consequences of an event or activity, and determining the corresponding risk. The risk is determined qualitatively in the early stages of a project, or quantitatively in the later stages of the project when more site-specific data are available as (Task 3 in Figure 5.1). Once the hazards and consequences have been identified, then the risk can be evaluated to decide whether or not it is acceptable (Tasks 4 and 5 in Figure 5.1). If the risk is acceptable, then the project can proceed to construction, or if the risk is unacceptable, then either the hazards and/or consequences are modified to reduce the risk to an acceptable level (Task 6 in Figure 5.1), or the project may need to be abandoned if the risk is too high.

This chapter discusses the criteria that are used to evaluate risk and the conditions under which the criteria are applied. It is beneficial to have well established and acceptable risk evaluation criteria so that different projects can be consistently evaluated. Risk evaluation criteria may be those that are widely used such as [F – N] diagram (see Section 5.2 below), or corporate standards developed for specific industries (see Section 5.3 below).

In addition to commonly used risk evaluation criteria discussed in Sections 5.2 and 5.3 below, it is possible to use event trees and decision analysis to examine the risk components of a project and then objectively decide on the suitable course of action (see Section 5.4 below). The event and decision trees identify the probability of occurrence of each chance event and its consequence and also incorporate the costs of remedial work. These analyses allow alternative courses of action to be evaluated, and then the most favourable decision to be selected.

5.1 RISK CALCULATION

Risk calculation usually involves developing a risk matrix that shows the likelihood of occurrence of an event, as well as the consequences of the event (see also Figure 4.2). That is, if the likelihood of occurrence is "almost certain" or "very likely", and the expected consequence is "severe" or "catastrophic", then the risk will be very high ("VH") and remedial action would

DOI: 10.1201/9781003271864-5

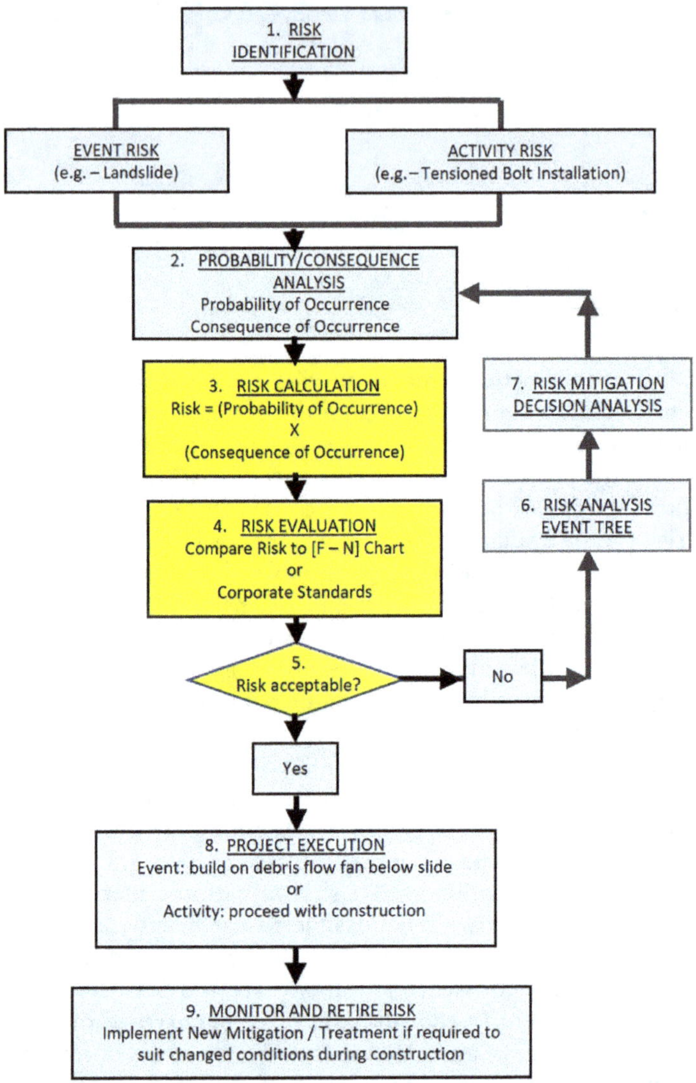

Figure 5.1 Flow chart of risk management tasks - Tasks 4 and 5, risk evaluation and acceptance.

usually be required. However, for an "unlikely" or "very unlikely" occurrence that has a possible "severe" consequence, the risk may be medium ("M") or low ("L"). That is, the risk of very rare events is usually tolerable by society, especially if they are involuntary such as natural disasters. The risk matrix readily allows the influence of different values of event occurrence and consequence to be evaluated.

Table 5.1 shows a risk matrix for possible project delays, and the corresponding consequences and costs for the delays. In the matrix, the probabilities (*P*)

Table 5.1 Typical risk management matrix for geotechnical projects (RoadEx Network, 2019)

Probability (P)		*CONSEQUENCES (can be amended to suit contract circumstances)* *Either TIME dependent*	*or COST dependent*	*Consequence (C)*		*RISK* R = (P · C)	*Degree of risk*	*Suggested action*
Very Likely >75%	5	>10 weeks added to planned completion date	>$1M	Very high	5	17–25	Not acceptable	If risk cannot be reduced project should not proceed
Likely 50–75%	4	>4 weeks added to planned completion date	$100k to $1M	High	4	13–16	Not acceptable	Work must not start until risk has been reduced
Probable 25–50%	3	>4weeks<1wk added to completion date	$10k to $100k	Medium	3	9–12	Significant	Reduce risk (mitigate or transfer)
Unlikely 10–25%	2	1 to 4 weeks on activity: no change to planned completion date	$1k to $10k	Low	2	5–8	Tolerable	Consider risk reduction measures
Negligible <10%	1	<1 week to activity: no change to planned completion date	<$1000	Very low	1	1–4	Trivial	Monitor work

and consequences (C) are given descriptive terms such as "probable" occurrence and "medium" consequence, as well as a corresponding numeric score of 1 through 5. The calculated risk is the product of the probability and consequence scores, and risk is ranked from "Trivial" (risk score 1–4) to "Not acceptable" (risk score 17–25). For each of the risk categories a corresponding action is suggested ranging from "Monitor work" for Trivial risk, to "if risk cannot be reduced, project should not proceed" for risk that is Not acceptable risk. The Suggested Actions also include references to mitigation work that can be undertaken to reduce the Probability and/or the Consequence in order to achieve a Tolerable or Trivial Risk.

It is noted that the risk matrix shown in Table 5.1 applies to *activities* such as components of a construction project that could result in delays, and not to *events* such as landslides that may occur repeatedly but at unpredictable times. A typical risk matrix for events is shown in Figure 4.2.

5.2 EVENT RISK – COMPARE TO [F – N] DIAGRAM

For event type risks such as landslides, debris flows, and rock falls where their occurrence is infrequent and unpredictable, the acceptability of the level of risk can be assessed from the [F – N] diagram as shown in Figure 5.2. The [F – N] diagram relates, on the vertical axis, the average annual frequency, F of events that could cause fatalities to, on the horizontal axis, the potential number of fatalities, N resulting from an event. The body of the chart defines four regions related to risk acceptability as follows.

5.2.1 Unacceptable risk

Unacceptable risks are relatively high-frequency events, with a lower bound of once every 1,000 (1.0E-3) years for events resulting in a single fatality, to once every million years (1.0E-6) for events that could result in 1000 fatalities. For landslides where the risk is in the Unacceptable category, a common mitigation measure would be to construct a barrier or fence, suitable to the conditions, to reduce the consequence of the event in terms of fatalities or property damage. The alternative mitigation strategy, if appropriate for the conditions, would be to work in the source area to improve the factor of safety and reduce the frequency of events. For example, rock anchors could be installed to stabilize rock slopes, or drainage measures could be implemented to control surface and groundwater pressure. If neither mitigation strategy is feasible nor economic, the project may need to be abandoned.

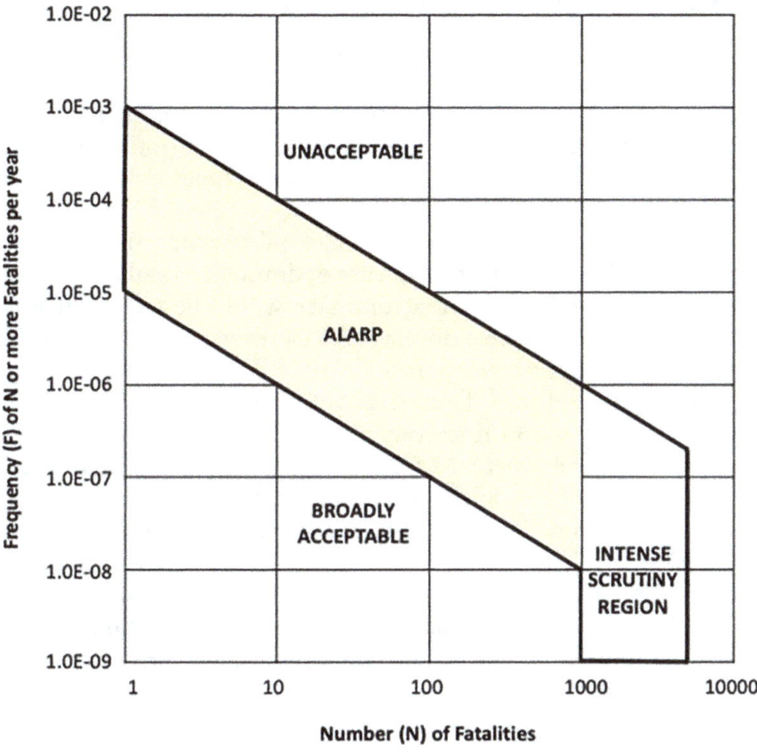

Figure 5.2 [F - N] diagram for evaluating the acceptance of event risk.

5.2.2 Broadly acceptable risk

Events are broadly acceptable if the frequency of occurrence is less than once in 100,000 (1.0E-5) for a single potential fatality, and the accepted frequency is one in 100 million years (1.0E-8) for up to 1,000 fatalities. This acceptable range of event frequency is consistent with the society's general acceptance of very low-frequency events, even if they are of high consequence, that people are very unlikely to experience in their lifetime. In general, society is accepting of rare events if they are somewhat familiar, are of natural origin and receive little media attention. Also, risk is acceptable if the benefit is perceived to outweigh the risk, especially if the cost of mitigation is very high. An example of an acceptable, very low frequency/very high consequence, and involuntary, event, is an asteroid impact with an urban area of the earth (Gardner, 2009). However, the successful 2022 DART (Double Asteroid Redirection Test) test to change the orbit of an asteroid shows that even this risk can be mitigated.

5.2.3 ALARP (as low as reasonably possible)

Risks that are intermediate between Unacceptable and Broadly Acceptable risks are those classified as ALARP. ALARP risks are those where the cost of mitigation is considered to be disproportionate to the benefits gained. That is, no mitigation could be performed, or a mitigation program could be carried out to reduce the risk, but a level of residual risk would remain that is considered to be acceptable.

For example, mitigation measures that would reduce the residual risk within the ALARP classification, is the use of drainage to stabilize landslides. If study of the landslide shows that drainage would be an effective means of improving stability, possible drainage measures would be to either drill a series of drain holes in the lower part of the slide, or, at much greater expense, drive a drainage tunnel into the base of the slide. Examples of severe consequence slides are those in reservoirs above the Clyde Dam on the Clutha River in New Zealand where 14.5 km of drainage tunnels were driven in seven slides (Macfarlane & Silvester, 2019), and the Downie Slide above the Revelstoke Dam on the Columbia River in Canada where 2.45 km of drainage tunnels and 24,000 m of drain holes were drilled (Kalenchuk, Hutchinson, & Diederichs, 2009). In these two cases, the consequence of sudden collapse of the slides into the reservoirs could be a large number of fatalities and very significant economic loss. In order to maintain the landslide risk at an acceptable level, the slides have been drained to lower the water table by driving, and maintaining, drainage tunnels below the slides. The cost of the adits and drain holes was justified by the high consequence of instability.

5.2.4 Intense scrutiny region

For situations where the potential number of fatalities exceeds 1,000, the risks and required mitigation measures would need intense scrutiny, even if

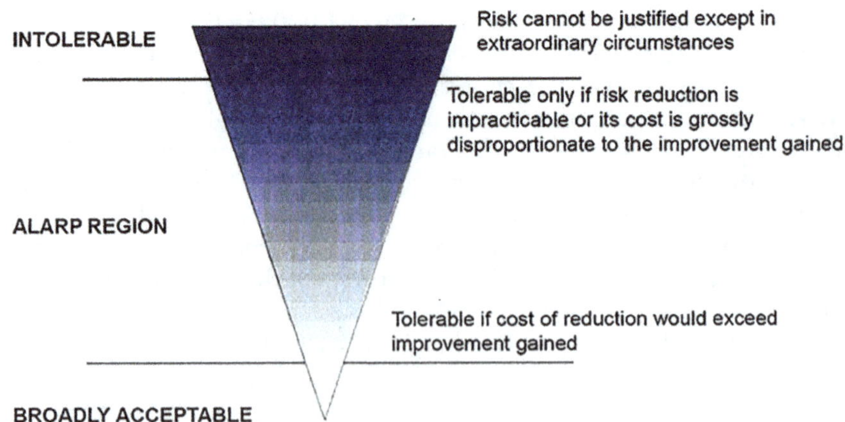

Figure 5.3 Illustration of relationship between classifications of event risk acceptance and required mitigation costs (ISO, 2019).

the frequency of occurrence is less than once in 1 million years. An example of projects where the potential number of fatalities may exceed 1000 are dams constructed in valleys upstream of population centres (see Figure 1.5, Kariba Dam).

Figure 5.3 is a graphical illustration of the relationship between the risk acceptance classifications. The inverted triangle indicates that required mitigation costs for events that are close to the Intolerable/ALARP boundary are greater than the mitigation costs for events that are close to the Acceptable/ALARP boundary.

Figure 5.4 provides a perspective of how the risk classifications in Figure 5.2 relate to many types of projects and common activities, and extends the [F – N] diagram to include high-frequency events (every year to once every 100 years, 1E+00 to 1E-02), but low hazard events that result in no fatalities (Read & Stacey, 2009). For example, Figure 5.4 shows the design criterion for open pit slope design. In open pit mines the consequences of a slope failure are mainly economic, with fatalities being unlikely because mining operations are carefully controlled and slope movement monitoring can usually provide a warning of deteriorating stability conditions that allow the site to be evacuated before failure occurs. Figure 5.4 shows that the Recommended Design Criterion for pit slopes is for an annual probability of occurrence that is about an order of magnitude lower probability than actual events showing that the current frequency of pit slope failures may not be acceptable, based on economics.

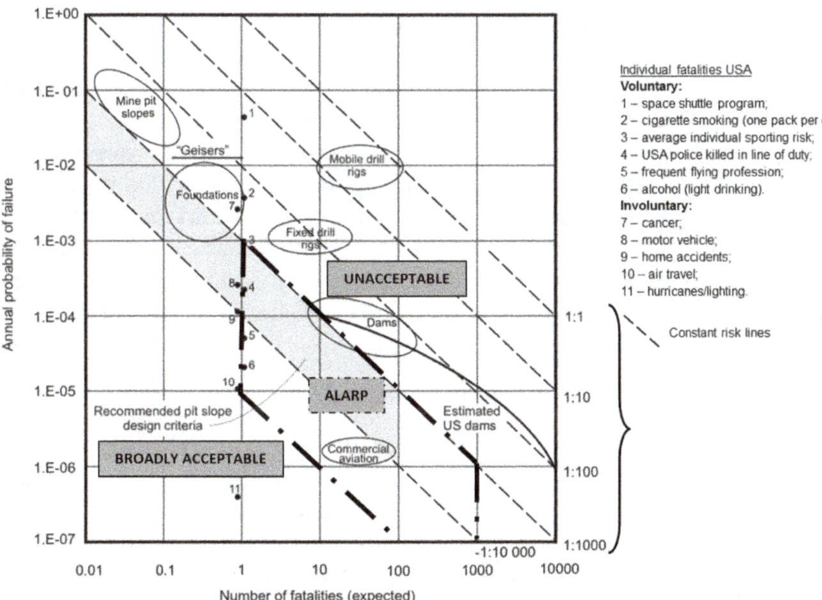

Figure 5.4 Frequency (F) - number of fatalities (N) relationships for common projects and activities showing acceptable, unacceptable and ALARP zones (see also Figure 5.2) (Steffan, Terbrugge, Wesseloo, & Venter, 2006).

Table 5.2 Suggested individual risk levels for landslides (Duzgun & Lacasse, 2005)

Slopes	*Individual Risk (Loss of Life/Year)*
Natural slopes	10^{-3}
Existing engineered slopes	10^{-4} to 10^{-6}
New engineered slopes	10^{-5} to 10^{-6}
Existing slopes	10^{-4}
New slopes	10^{-5}

Figure 5.4 also identifies risk conditions for foundations where settlement resulting in unacceptable performance of a structure occurs every 100–1,000 years, but is unlikely to result in fatalities. It is highly likely that the initial settlement of a building will provide an early warning of deteriorating conditions allowing for timely evacuation; the exception would be settlement and collapse caused by an earthquake. Excessive foundation settlement that occurs with an annual frequency of more than about 1E-02 (100 years) is generally unacceptable for economic reasons.

Figure 5.4 shows the difference between voluntary activities such as flying in the Space Shuttle and smoking, and involuntary risk such as cancer and motor vehicle accidents. In general, voluntary risks for activities in which individuals choose to participate, are more acceptable to society than involuntary risks.

Figure 5.4 shows a curve of estimated USA dam failures for data in a report on life safety consequences of dam failures in the USA (Stanford University, 2011). The report identifies 63 dam failures resulting in fatalities over a 167 year period between 1850 and 2016, at an annual rate of 0.34 failures per year. Of the 63 failures, only 10 (16% of total) were the result of failure of the embankment or foundation, a geotechnical condition, and most of the other failures were due to overtopping resulting from floods or spillway operations.

Further guidance on acceptable risk for landslides is provided by work in Australia, Hong Kong and Canada where the voluntary and involuntary risk is recognized – voluntary risk refers to natural slopes, and involuntary risk refers to engineered slopes. The suggested acceptable individual annual risk levels (loss of life per year) for landslides are shown in Table 5.2.

5.2.5 Example of event risk – talus slope

Accumulation of rock falls to form a talus slope, and the study of the hazard to activities on or close to the base of a talus slope is an example of event risk analysis. In glacial areas of Canada talus has accumulated since the retreat of the glaciers about 10,000 years ago so it is possible to develop a probability distribution for the rate of accumulation from estimates of the number and sizes of the blocks of rock in talus, and a probability distribution for the possible time of accumulation. As discussed in Section 2.4.3, for a talus slope at the base of a 320 m high by 750 m long rock face that has accumulated over a period of about 10,000 years, the average number of rock falls is about three falls per year, with a range of 1–15 falls per year.

The hazard of rock falls to personnel working at the base of the talus depends on the spatial and temporal situation of the work. That is, how much of the 750 m long slope is occupied by the work activities, and how much time, such as the number of days per year, is the work active. If the work occupies only 2% of the site (length 15 m) and is active for 5% of the year (18 days/year), then the annual probability of personnel being on site when a rock fall occurs is: $P = (0.02 \cdot 0.05 = 0.001)$ or 1E-03. If five personnel are at risk during this work (number of fatalities. N = 5), Figure 5.4 shows that the risk is just in the Unacceptable zone (above ALARP zone). Another probability issue that would be taken into account in the analysis is the fact that only a very small number of rocks falls roll (~0.5%) beyond the lower limit of the talus so the actual probability of falls that may impact the work personnel is about 1E-03 × 0.005 = 5.0E-6, or Broadly Acceptable.

5.3 ACTIVITY RISK

The use of the [F – N] diagram in Figures 5.2 and 5.4 mainly relates to events such as landslides that occur repeatedly, but at uncertain intervals, for which an annual frequency of occurrence can be estimated or calculated, depending on available records of previous events. However, risks also exist for activities that are carried out once but can result in unsatisfactory outcomes that are usually economic such as cost overruns and project delays, rather than the cause of fatalities. For these conditions, a different set of criteria are required to assess the Acceptability/Unacceptability of a project, and different methods of mitigation are necessary, as discussed below. This type of risk may be termed Activity Risk.

The procedure to quantify incident risk and determine an appropriate management response is shown in Figure 5.5 and Tables 5.3 and 5.4.

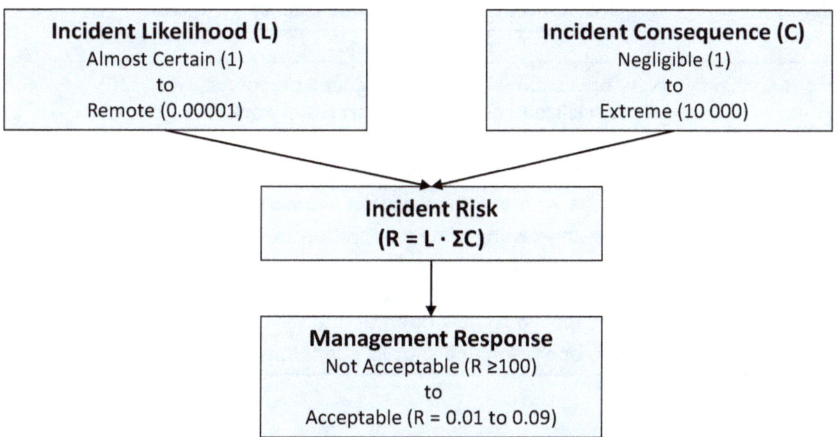

Figure 5.5 Structure of activity risk calculation and management response.

5.3.1 Descriptions of likelihoods and consequences for activity risk

For Activity Risk it is possible to assign numeric scores to both the Likelihood of the activities occurring, and their Consequences. The product of these two scores is the risk score for the activity, and this score can be compared to established corporate Acceptance risk scores. For example, the risk score can be used to select one of three courses of action: (a) accept that the risk is acceptable; (b) mitigation is required to reduce risk to an acceptable level; (c) the risk is unacceptable and the project should be significantly modified or even abandoned. Tables 5.3 and 5.4 provide respectively, descriptions of the Likelihood and Consequence categories, and their corresponding numeric scores.

Having determined scores for both the incident Likelihood, L (Table 5.3) and the incident Consequence, C (Table 5.4), the resulting Risk score can be calculated from the product:

Risk, $R = [(\text{Likelihood}, L) \cdot (\text{Consequence}, \Sigma C)]$

The term ΣC refers to the sum of all the consequences that may result from the event. That is, the consequences of a slope instability incident on a highway could be could be injuries, damage, and traffic interruptions, for each of which a score would be assigned, with the total consequence score being the sum of the three scores.

For the calculated Risk score, the corresponding management response can be selected from Table 5.5.

For Activity-type risks, it is necessary to customize The Likelihood and Consequence descriptions listed in Tables 5.3 and 5.4 for specific industry and corporate situations.

Table 5.3 List of incident Likelihood descriptions, and corresponding numeric scores

Incident likelihood (L)		
Likelihood Category	*Conditions related to likelihood descriptions (Must meet* at least one *of the criteria)* *Activity description*	*Likelihood score, L*
Almost Certain	New operation with no experienced personnel or specialized equipment to carry out work	1
Probable	Personnel with required equipment have some experience on similar applications	0.1
Possible	Specialists with experience assist site personnel	0.01
Unlikely	Training and testing of new operation carried out before starting work	0.001
Rare	Good experience with similar operations that can be applied to new operation	0.0001
Remote	Routine operation under usual conditions	0.00001

Table 5.4 List of incident consequences and corresponding numeric scores

	Incident consequence (C)				
Consequence sub-category	*Negligible*	*Low*	*Medium*	*High*	*Extreme*
Health/safety: • Public • Employee • Contractor	Incident causes no Health and Safety consequence.	Minor injury or minimal consequence to person, possibly result in a small claims court settlement.	Serious injury, or adverse short-term health consequence, recovery with no long-term disability.	Multiple serious injuries, or one fatality, or adverse long-term health consequence.	Multiple fatalities
Property damage to • Public • Private	Incident causes no/negligible damage to public property.	Minor repair required to facilities.	Larger scale damage with possible associated fire.	Destruction of a single dwelling, small building, or similar small structures.	Destruction of multiple dwellings, major industrial installations.
Environmental: • Emissions; • Chemical discharge; • Damage to flora/ fauna/ habitat	Incident with no/ negligible environmental consequences.	Incident resulting in minor public/ regulatory concerns.	Event resulting in moderate environmental consequences.	Long-term or major environmental damage (may include media involvement).	Uncontrolled environmental damage (media will be involved).

(*Continued*)

Table 5.4 (Continued) List of incident consequences and corresponding numeric scores

Social: • Societal • Corporate image	Incident causes negligible or minor public disturbance.	Minor incident causing inconvenience to public.	Incident causes delays to traffic of several hours	Incident delays traffic for up to one day, minor damage to nearby structures, negative publicity.	Incident causes traffic delays of several days, damage to nearby structures needs urgent repair, widespread publicity.
Government relations: • Municipal • State • Federal	Incident causes no external relations issues.	Incident leads to communication and/or notification to external agencies, minimal follow-up required.	Incident leads to communication with local municipality, with follow-up discussions, reporting required.	Discussions with government agencies, requiring increased oversight of construction, possible modifications to methods.	Government agencies require modifications to construction procedures that will have impact on current and future operations.
Project cost/delay impact	<\$1,000	\$1,000–\$5,000	\$5,000–\$50,000	\$50,000–\$500,000	>\$500,000
Consequence score, C =	1	10	100	1,000	10,000

Table 5.5 List of risk scores (R) calculated from likelihood (L) and summed consequence (C) scores and corresponding management response - ($R = L \cdot \Sigma C$)

Incident risk score and response		
Risk level	*Management response*	*Score $R = (L \cdot \Sigma(C))$*
I	Not Acceptable - Notify Management, consider alternate development	≥100
II	Not Acceptable - Notify Management with Mitigation Strategies	10–99
III	Discuss with Management	0–9
IV	Acceptable - Follow-up required	0.1–0.9
V	Acceptable	0.01–0.09

For example, the Likelihood descriptions listed in Table 5.3 relate to a construction project risk where the experience of the contractor for a specific activity is being studied. A similar table could be developed for risks related to the uncertainty of subsurface conditions resulting in possible negative outcomes due to instability of excavated slopes, or of groundwater inflow and disposal.

Regarding consequences of negative outcomes, customized consequence lists are also required. For example, for utility companies (e.g., gas and electrical supply) the consequences will be closely related to reliable supply to their customers and the duration of outages, as well as their relationship with government regulatory agencies that are tasked with monitoring the reliability and safety of supply. Also, for highway systems that are usually government owned and operated, the consequences will be related to service disruptions due to landslides, for example, that cause traffic delays, or poor maintenance that results in accidents. For railways that may be run by private companies, consequences relate to on-time performance, traffic delays and safe operations, with somewhat different consequences for freight trains, and for passenger trains where consequences could be injuries and fatalities. The importance of schedules for freight transport is that the trains are often a vital part of supply chains linking manufacturing to consumers.

The examples of management response listed in Table 5.5 provide a guideline on a relationship between possible responses and calculated risk scores. It is likely that calibration of the Response/Risk relationship will be required to suit particular situations.

It is noted that the customization of Likelihood and Consequence descriptions for Activity risks in Tables 5.3 to 5.5 contrasts with the frequency/fatality relationship ([F – N]) shown in Figures 5.2 and 5.4 for event risks. One of the objectives for assessment of event risk is to apply a consistent risk approach to a range of different site conditions so that selection of Acceptable/ALARP/Unacceptable risk is applied as uniformly as possible. For example, it would be unsatisfactory if adjacent developments in similar, steep terrain were not assessed to the same risk acceptance standard.

5.3.2 Calculation of activity risk

Use of the numeric scores for Likelihood and Consequence shown in Tables 5.3 and 5.4 respectively, the following method is used to calculate risk from the two scores.

Table 5.3 shows the six categories of Likelihood, each with a corresponding score:

- Almost certain - score: 1
- Probable - score: 0.1
- Possible - score: 0.01
- Unlikely - score: 0.001
- Rare - score: 0.0001
- Remote - score: 0.00001

Application of Table 5.3 requires for each risk, selection of the Likelihood of its occurrence based on experience of previous and similar circumstances, and knowledge of site conditions. A score is assigned to the selected Likelihood, noting that the use of a logarithmic scale for the scores helps to clearly differentiate between each category of Likelihood.

Table 5.4 shows five categories of Consequence severity each with a corresponding score:

- Negligible - score: 1
- Low - score: 10
- Medium - score: 100
- High - score: 1,000
- Extreme - score: 10,000

Table 5.4 also lists five sub-categories of Consequence:

- Health and safety
- Property damage
- Environmental impact
- Government relations
- Cost

Application of Table 5.4 requires that, for each of the sub-categories of Consequence, one of the five levels of severity is selected, again based on experience and knowledge of site conditions.

5.3.3 Example of activity risk – highway construction

An illustration of Activity risk calculation and the corresponding management response is as follows. For a project involving the construction of a new interchange on an existing highway, the descriptions of the incident likelihoods that could result in negative outcomes such as traffic interruption, schedule extension and cost overruns are:

Almost certain – low strength, variable soils of uncertain extent require deep foundations of unknown depth (L=1)
Probable – limited investigation program confirms deep, low strength soils but allows foundation designs to be prepared (L=0.1)
Possible – investigation program defines limited areas of low strength soils allowing appropriate foundation designs to be prepared (L=0.01)
Unlikely – extensive investigation program defines subsurface conditions to be favourable for construction (L=0.001)
Rare – minor modifications required to previous designs (L=0.0001)
Remote – essentially identical construction to nearby interchange (L=0.00001)

If the site investigation shows that the extent of the low strength soils is limited so that the anticipated likelihood of negative outcomes can be classified as **Possible,** then the Likelihood score is **L=0.01.**

For the interchange construction project, the selected severity level for each Consequence sub-category is listed below.

Health and safety – **Low** severity, minor injury possible
Score=10
Property damage – **Negligible** severity, no private property at risk
Score=1
Environmental – **Negligible** severity, environmental impact study complete
Score=1
Social – **High** severity, schedule delays for foundations, traffic disruption possible
Score=1,000
Government relations – **Medium,** public consultation on consequences complete
Score=100
Cost – **High** severity, traffic delays and interruptions could increase construction time
Score=1,000

Total Consequence score, ΣC=(10+1+1+1000+100+1000)=2,112

For the selected Likelihood score of L=0.01 and the total consequence score of ΣC=2112, the calculated risk score is:

$R=(L \cdot \Sigma C)=(0.01 \cdot 2{,}112)=\mathbf{21.12}$

Table 5.5 shows that the management response for a risk score of 21.12 is **Not Acceptable** and that mitigation is required. A suitable mitigation strategy could be to modify the construction plan to reduce traffic interruptions so that the Social and Cost consequences scores are both reduced from 1000 to 10 for which the new total consequence score is 132, and the new Risk score is:

$R=0.01 \times 132 = \mathbf{1.32}$

The modified construction plan may be acceptable to management.

It is noted that the Activity Risk analysis discussed above in Section 5.3.3 illustrates a possible strategy for risk management for a project involving highway construction. Details of the Likelihood and Consequence descriptions and Management Responses should be customized to the requirements of each project.

5.4 EVENT TREES, DECISION ANALYSIS

As described in Section 5.3 above, determination of risk involves assessing the Likelihood of occurrence of an event and its Consequence, from which the risk is calculated. The next tasks are to make a decision if the risk is acceptable, and if not, to implement a mitigation plan to reduce the risk to an acceptable level.

One procedure that can be used to rationally select an optimum mitigation plan is to first develop an Event Tree defining the structure of the risk, and then use Decision Analysis in which the cost of the mitigation strategy is combined with the probability of event occurrence and the consequence for each strategy. For each mitigation option, the mitigation cost combined with the probabilities of event occurrence and consequence to calculate an Expected Value (EV) – the optimum strategy is the option with the least EV.

Event trees are discussed in Chapter 6, and Decision Analysis in Chapter 7.

5.5 CASE STUDIES RISK EVALUATION

The following is a discussion on the risk evaluation and acceptance for the four case studies.

5.5.1 Debris flow containment dam

Debris flows are a common hazard in mountainous terrain containing steep-gradient creeks where periods of intense rainfall can occur. Where urban development and transportation routes are located on these creeks, or on the run-out fans, catchment dams are often constructed to temporarily contain the debris material and protect downstream facilities. The hazards to these facilities are that the dam will be overtopped, or even fail in the case of an exceptionally large volume event. A related hazard is that a series of small events occur that partially fill the catch basin, but it is too dangerous to work in the catch basin and remove accumulated debris, resulting in the dam eventually being overtopped.

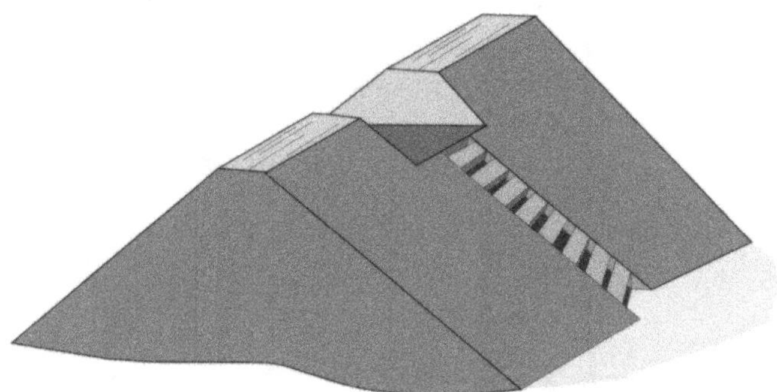

The required capacity of the debris flow dam and its catch basin is determined by the [F – N] relationship shown in Figure 5.2, and the [magnitude – frequency] relationship for debris flows at the site. That is, if the dam is located above an urban development where 100 people are at risk if the dam is over topped, then the [F – N] diagram shows that the acceptable annual frequency of such an event should be less than 1.0E-07, or that the annual frequency should be less than 1.0E-05 to be within the ALARP range.

Study of the stream channel upstream of the dam, the precipitation in the catchment area and the history of previous events will allow a relationship to be developed between the magnitude of future events and their frequency of occurrence (Guthrie & Evans, 2005). This information can be expressed as a [frequency – magnitude] curve that can be used to size the dam and its catchment basin volume to contain events at annual frequencies of between 1.0E-05 and 1.0E-07 for the risk of fatalities to be in the ALARP range.

An alternative mitigation strategy would be to ban development of permanent residences or transportation facilities on the run-out fan. Selection of the optimum strategy would depend on the relative costs and reliability of constructing a containment dam of the required capacity, compared with the value of the land on the run-out fan (see Section 7.6.1 below for the use of decision analysis to evaluate debris flow protection options).

5.5.2 Rock fall hazard

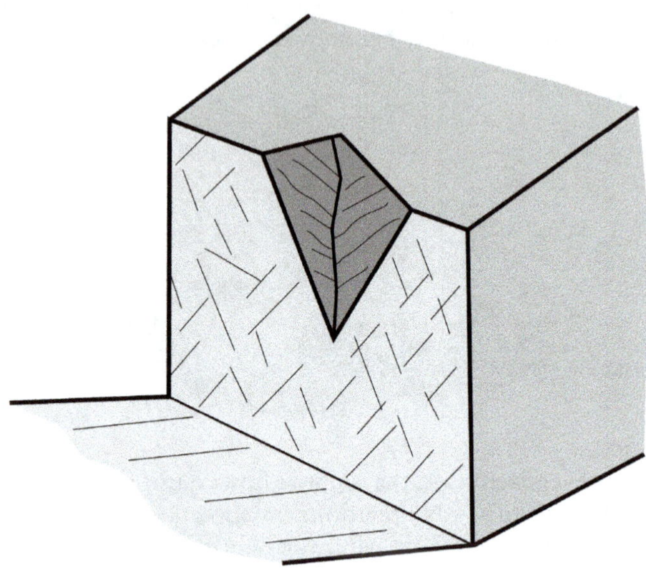

In mountainous terrain, many linear systems such as pipelines, railways and highways, may be at risk from slope instability hazards. The severity of the hazard will depend on the frequency and magnitude of the slides, and the consequence of slides in terms of disruptions to traffic, damage to track or equipment, and injury or fatalities to personnel or the public.

Where these systems are a vital component of the economy, and where the public is exposed to the hazards, a risk management program is often implemented with the objective of decreasing the frequency and magnitude of the events by implementing stabilization programs, or controlling the consequences by installing protection measures such as fences or sheds.

Risk management programs for transportation systems often comprise the following components:

- Inventory of geotechnical features such as slopes, bridge foundations and tunnels, that quantifies the risk in terms of both the potential hazards and consequences.
- Records of incidents such as rock and soil slides, and their consequence such as traffic delays.
- Proactive stabilization program in which work focuses on the highest risk locations as identified in the inventory; stabilization programs are usually ongoing because of the progressive weathering and deterioration of the rock and soil, and the effects on stability of rain and freezing.

The following is an example of an inventory system for slopes along a highway system (Wyllie, 1987); (Pierson, Davies, & Van Vickle, 1990).

The slope inventory is divided into two classes:

- **Hazards** of the slope (physical characteristics) – height, geology, block size, weather, and rock fall history.
- **Consequences** of a fall (operational characteristics) –ditch effectiveness, average vehicle risk (AVR), percentage of decision sight distance and roadway width.

For each of these nine characteristics, a numerical score is assigned that ranges from 3 for favourable conditions to 81 for very unfavourable conditions. Intermediate conditions are assigned scores of 9 and 27 points, with the geometric increase in scores from 3 to 81 being used to clearly distinguish between very low-risk and very high-risk slopes. The total hazard score is the sum of the five scores for the physical characteristics, and the total consequence score is the sum of the four scores for the operational scores.

The final risk score for a slope will be the product of the total of the five hazard scores and the total of the four consequence scores. This scoring system clearly recognizes, for example, the influence of remedial work such as a significant improvement in the effectiveness of the ditch in containing rock falls that reduces the ditch effectiveness score from 81 to 3.

As an example of the slope hazard scoring system, consider the following situation:

- Slope height of 32 m - score = 81
- Excavated in rock containing continuous, planar joints dipping out of the face - score = 27
- Blocks with volumes up to about 1 m^3 - score = 27
- Climate with moderate precipitation levels and short duration freeze-thaw periods - score = 9
- History of frequent rock falls - score = 27.

Total hazard score:

= [81 (height)+27 (geology)+27 (size)+9 (climate)+27 (frequency)]=171

For this slope, the consequences of a fall are defined by:

- Ditch with no significant catchment capacity - score = 81
- AVR of vehicles being present below the slope at 50% of the time - score=9
- Decision sight distance of 80% of the design value - score = 9
- Roadway width, including shoulders of 8.5 m that allows only limited avoidance action to be taken by the driver - score = 27.

Total consequence score:

= [81 (ditch effectiveness)+9 (AVR)+9 (sight distance)+27 (roadway width)] = 126

Risk score=(171 · 126)=21,546

If the ditch is deepened and widened such that the effectiveness is improved to a score of 3, then the new total consequence score is [(3+9+9+27)=48] and the new risk score is [171 · 48=8,208]. The new risk score is a 62% decrease in the total score clearly showing an improvement in the rock fall risk conditions.

The inventory will identify the highest risk slopes from which a proactive stabilization program can be developed in which the work can be prioritized according to the total risk score.

In addition to the slope inventory, detailed records can be maintained of instability events that include location, date, magnitude, preceding weather conditions and geology, and the consequences of the event such as traffic interruptions; the records would also include the stabilization work. These records of instability events and stabilization can be used to update the inventory scores, so that, for example, if rock fall frequency increases significantly due to rock weathering, then the frequency score could increase from 27 to 81.

Regarding the acceptance of rock fall risk expressed by the inventory score, this cannot be related directly to the [F -N] diagram shown in Figure 5.2 because the hazard can still be significant without any fatalities having occurred. A possible approach to accepting the risk of the system is to determine a total risk score that is acceptable to management, consistent with all other hazards such as pavement condition and guard barriers, and then to implement a stabilization program with the objective of reducing the score of all slopes to below the acceptance level.

5.5.3 Tunnel stability

One of the possible hazards that can occur when driving tunnels in rock is that the excavation will intersect zones of weak and closely fractured rock that may contain flowing water. When this condition occurs unexpectedly, rock falls, water inflow and possible tunnel collapse can be a hazard to workers, and result in schedule delays and extra costs to stabilize the tunnel.

A possible risk management approach to this situation is to drill probe holes ahead of the face to obtain information on rock conditions. This information can be used to plan and carry out remedial work prior to advancing the tunnel. Remedial work could include grouting to consolidate loose, flowing rock, or installing spiling to create an arch of reinforced rock above the tunnel (Hoek, 1980).

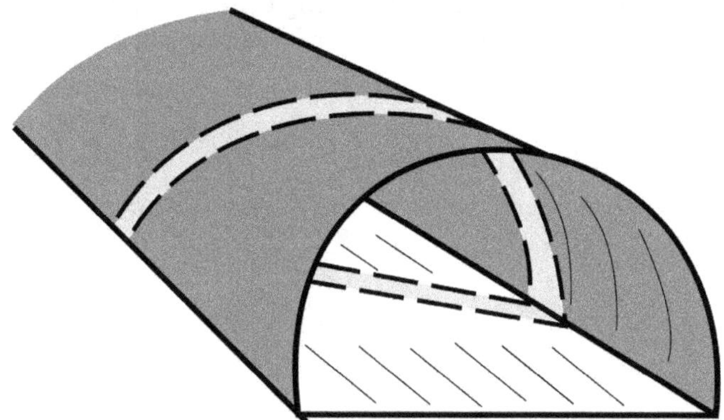

The decision to drill probe holes ahead of the face will depend on how well the geology along the tunnel alignment is known, and whether the cost and time of drilling the holes is justified by the reduction in risk of intersecting poor ground with little warning. A decision analysis to compare these two options is discussed in Chapter 7, Section 7.6.3.

Regarding the acceptance of the risk of tunnel instability, these conditions are not applicable to the [F – N] diagram in Figure 5.2 that applies to recurring events that occur at an annual frequency (F) whereas the tunnel stability risk is a single event. Therefore, the acceptance of risk and the decision to drill probe holes ahead of the face would have to be based on a combination of the tunnel support design, and the experience and judgement of the tunnel foreman and engineer.

A commonly used method to design tunnel support that is applicable to rock conditions is the Barton Q system in which the quality of the rock is expressed in terms of six rock properties, the values of which are used to calculate the rock mass quality parameter, *Q* (Barton, Lien, & Lunde, 1974):

$$Q = \frac{RQD}{J_n} \cdot \frac{J_r}{J_a} \cdot \frac{J_w}{SRF}$$

where *RQD* is the Rock Quality Designation, J_n is the joint set number, J_r is the joint roughness number, J_a is the joint alteration number, J_w is the joint water reduction factor and *SRF* is the stress reduction factor. The *Q* value, the tunnel span and the importance/longevity of the excavation (*ESR*) are related to the appropriate support design. If the probe holes identify rock with a low *Q* value, then steps could be taken to reinforce the rock ahead of the face. See Section 2.8.3 for an example of calculating the probability distribution for *Q* that takes into account uncertainty in the parameters defining *Q*.

Use of a well proven method of designing tunnel support will be of assistance in making a decision on the need for drilling probe holes that meets the acceptable risk level for the project.

5.5.4 Dam foundation

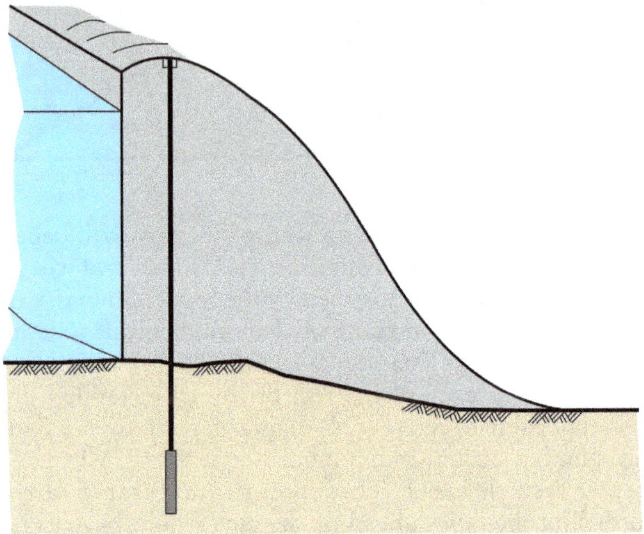

For many concrete gravity and arch dams in seismic areas constructed before about 1960, the ground motion design criteria have been significantly upgraded since the time of the original design. A common method of upgrading the stability of dams to meet the new design standards has been to install rock anchors that are secured in the dam foundation and then tensioned against the top surface of the structure. The tensioned anchors increase the normal force on the rock-concrete surface on the base of the dam to increase the resistance to shear displacement, and increase the overturning resistance.

Installation of tensioned rock anchors requires drilling carefully aligned holes through the dam into the foundation, inserting the anchor, grouting the bond zone in the foundation and then using a hydraulic jack to apply a tension load to the anchor. For substantial dams, the anchors are usually multi-strand cables and the applied tension load can be as high as 2,500 kN, and hundreds of anchors may be installed.

Although installation of tensioned anchors is usually a routine operation, installation in an operating dam requires attention to detail and strict quality control, to avoid, for example, the holes deviating into the upstream face of the dam. Planning for such an installation may involve deciding whether the cost of installing test anchor(s), with the objective of modifying and refining the construction procedures, is justified for achieving improved quality and potential savings in construction costs. A key procedure in installing acceptable tensioned anchors is to carry out cyclic, load–elongation tests in which the load–elongation relationships must meet specific acceptance criteria (Post Tensioning Institute, 2014).

The use of Decision Analysis to study the benefits of carrying out a test anchor program is illustrated in Section 7.6.4 below.

Chapter 6

Event trees – structure of risk events

An event tree is a graphical representation of the chain of events that may result from an initiating event (Baecher, & Christian, 2003). The result of the events may be a system failure such as a collapse of a tunnel or an unfavourable outcome of a project such as delays and cost overruns. The purpose of the tree is to show progressively, the relationship between events for each initiating event, the probabilities of occurrence of each event and the possible consequences that may result from the events.

In the overall structure of risk management, event tree analysis may be Task 6 in which the probability of occurrence of each hazard is examined to find which consequence of the hazards is most likely to occur (Figure 6.1). This information can be used to prioritize remedial work.

The event tree structure can be imported into decision analysis in which probability distributions for the costs of remedial work and the consequences of events can be analyzed to determine an appropriate remedial action (Canadian Dam Association, 2013). Decision analysis is discussed in Chapter 7.

An alternate representation of an event tree is an *influence diagram* that graphs relationships between initiating events, states of nature, conditions of the system and consequences (Baecher & Christian, 2003).

In addition to event trees that examine causes and the resulting consequences, the reliability of systems can be analyzed by *fault trees* which begin with the final system failure and reason backwards to identify causes of failure, such that the logic moves from consequence to cause. Because of the different logic between event trees and fault trees, they cannot be interchanged. Fault trees are commonly used to analyze mechanical systems such as power plants, but can also be applied to dams (McCann, 2002).

6.1 FEATURES OF EVENT TREES

Figure 6.2 shows a simple event tree comprising an initiating event (I) that can result in one of two possible events for system state 1: either a favourable

DOI: 10.1201/9781003271864-6

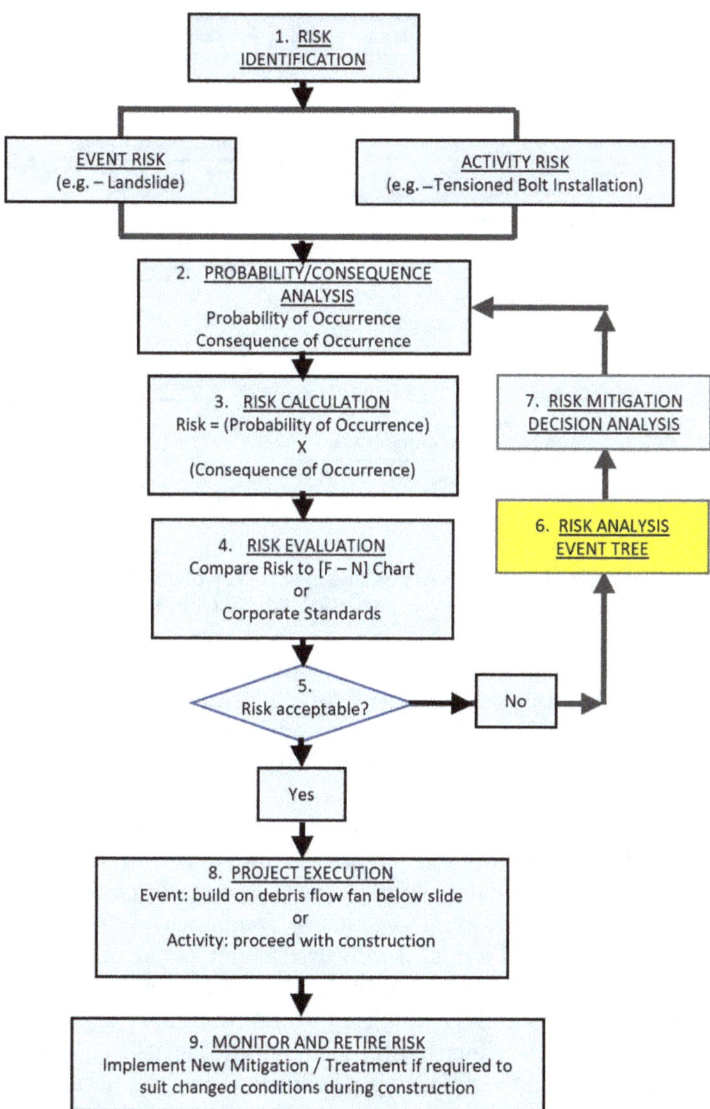

Figure 6.1 Structure of risk management program – Task 6 risk analysis.

event (success, S_1) or an unfavourable event (failure, F_1). Following each of the S_1 and F_1 events, additional success and failure events can occur for system state 2 such that final sequences of each branch are defined by the events along the branch such as: $[I \cdot S_1 \cdot S_2]$ for the two favourable events.

The usual terminology for event trees, as shown in Figures 6.2 and 6.3, comprises an "initiating event" such as an earthquake, and then all possible

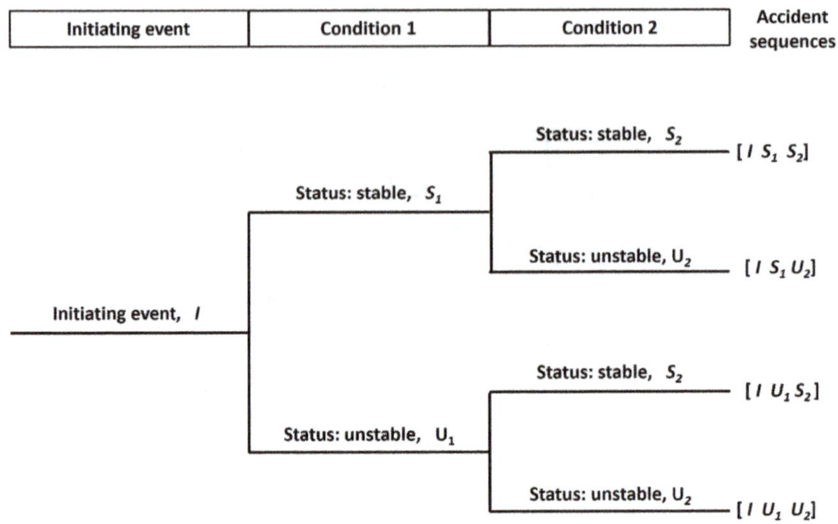

Figure 6.2 Generic event tree for two possible events resulting from the initiating event (U. S. Nuclear Regulatory Commission (WASH-1400), 1975). PPP.

events that may occur as a result of the earthquake such as a slope failure (Branch 1), possibly followed by closure of a highway (Branch 2). Each of the events is defined by a "branch", with a "node" as the issuing point for each chance event. Nodes may also be called "chance" points (shown as circles) because they represent the occurrence of uncertain events such as an earthquake or flood where the occurrence is quantified by assigning each event a probability value. The events on each branch have uncertain occurrences that are expressed as conditional probabilities (P). Each node can have any number of branches that are mutually exclusive events, each defined by its own conditional probability. That is, if two events may occur at a branch point and the probability of the event on Branch 1 is ($P1$), then the probability of Branch 2 event occurring is ($1 - P1$), and the sum of all the probabilities at each node equals 1.0 showing that these are all the events that can occur.

While each node can have many possible events, the actual event will only follow one path, and the probability of this event occurring is defined by the joint (or path) probability along that sequence of events. For each event, the joint probability is the product of the probabilities along that path. For example, in Figure 6.3, an initiating event that results in subsequent events that follow Branch 1 and then Branch 4 will have a joint probability of $[P1 \cdot (1 - P3)]$.

The joint probabilities for each of the four paths in Figure 6.3 are shown at the ends of each branch (shown as squares). It can be verified that the sum of these four probabilities is equal to 1.0, confirming that these are all the events that can occur as a result of the initiating event.

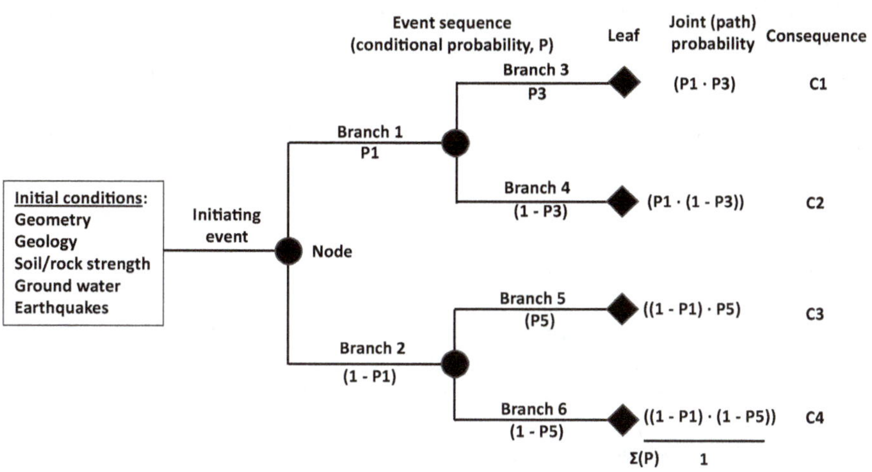

Figure 6.3 Event tree terminology showing sequence of events, conditional and joint probabilities, and consequences of events.

The termination of each branch is a "leaf" where the "consequence" of each branch is defined, such as interruptions to traffic and for repair to the highway, which are usually expressed as costs. In Figure 6.3, four consequences (*C1* to *C4*) are listed representing the expected consequences for each branch, with the joint (or path) probability showing the probability of occurrence of each. Numerical values for the probabilities and the consequences would indicate which consequence is the most likely to occur, and the potential magnitude of the damages. An event tree with calculated joint probabilities is shown in Figure 6.7.

It is usual to show the events in chronological order such as the occurrence of the earthquake followed by the landslide, although this is not necessary mathematically. Also, success states are usually shown above failure states.

The application of probabilities is discussed further in Section 6.4, and in Chapter 7 on Decision Analysis.

6.2 INITIAL CONDITIONS

The initial conditions for most systems will be internal conditions, such as foundation geology and soil/rock strength that define the system itself, and external conditions that are applied to the system, such as the water force on the dam and the mass of the dam structure both of which are defined by specific values. For slope stability studies, external conditions will be the height and face angle of the slope, as well as any stabilization work that has been installed such as tie-back anchors.

Regarding uncertainty in the initial conditions, external conditions such as the mass of the dam and the slope geometry, are usually known and have specific values. However, for less easily defined and variable internal conditions such as the soil and rock strengths, groundwater level and earthquake ground motions, their occurrence will be defined by probability distributions that would be established by the field investigation program.

6.3 INITIATING EVENT

Having defined the existing conditions of the system, the next component of the event tree is the initiating event that introduces loading on the system; events that are subsequent to the initiation, are the response of the system to loading. The type of initiating event will depend on whether the hazard is, (a) an event such as slope instability, or (b) an activity such as constructing a stable tunnel, as discussed below.

6.3.1 Event risk

Typical initiating events are earthquakes, floods or human error that occur randomly in time, and it is necessary to estimate the annual probability of occurrence of the event, or the probability during a defined time interval such as the life of the project. These random initiating events can be modelled by the Poisson distribution that is used to model the probability of occurrence of discrete events where independent events (x) are occurring at different times, over a fixed interval of time (t) and have a known constant mean rate, μ (Kreyszig, 2011).

The Poisson distribution for a single time period, such as one year, is given by equation (6.1):

$$f(x) = \frac{\mu^x}{x!} e^{-\mu} \tag{6.1}$$

where μ is the mean of the annual occurrence frequency that is constant over time, e is the base of the natural logarithm (value=2.7183) and x is the number of events.

For a time interval t, the number of occurrences in this time interval is (equation 6.2):

$$f(x) = \frac{(\mu \cdot t)^x}{x!} e^{-(\mu . t)} \tag{6.2}$$

An example of the application of a Poisson distribution to the risk of slope failure is as follows. Consider a series of slopes where records show that instability is often initiated when intense rainfall totalling at least 300 mm occurs over a 72 hour period. Analysis of rainfall records shows that these events occur, on average, every 20 years, or at a mean annual rate of [$\mu=1/20=0.05$]. Applying equation (6.1), it can be found that the annual probability of one intense rainfall event ($x=1$) that may initiate slope instability is 4.8%, and the probability of no intense rainfall is 95.2%. In comparison, if intense rainfall occurs more frequently at once every five years, the mean annual rate of intense rainfall is [$\mu=1/5=0.2$], and the probability of an intense rainfall event in a year is 16.4%. These probability values can be entered directly into the initiating event for event trees such as Branch 1 [$P_{rainfall}>300$ mm] and Branch 2 [$P_{rainfall}<300$ mm] in Figure 6.3.

Equation (6.2) can be used to find the probability of an event occurrence over a specified time period. If the mean annual rate $\mu=0.05$, then over a period of 10 years, the probability of one intense rainfall event is 30.3%, and the probability of two events is 7.6%. In comparison, equation (6.1) shows that the probability of one intense rainfall event in a single year is 4.8%, and of two events in a single year is 0.1%.

An important property of initiating events is that they are considered to be independent because, for example, each flood would be independent, and the occurrence of a flood has no relationship to an earthquake. Independent initiating events require that a separate event tree be developed for each initiating event because the response of the system will be different for each.

6.3.2 Activity risk

Activity-type risks are initiated by single events, which is in contrast to event-type risks, that are initiated by repeated events that occur randomly in time. For the activity risks described in the case studies such as driving (constructing) a stable tunnel, or installing tensioned anchors in a dam foundation, the initiating event is simply the decision to proceed with the project once it has been determined that the risks of delays and cost overruns due to unforeseen circumstance during construction are acceptable. Methods of studying risk acceptance are discussed in Chapter 5.

6.4 PROBABILITY OF OCCURRENCE

Each node (or chance point) on an event tree shows uncertain events that can occur, where the uncertainty is quantified by assigning each event a probability of occurrence. By listing each event, with its probability of occurrence, separately, it is possible to ascertain the influence of each event on the consequences of the initiating event. The term conditional probability is a measure of the probability of an event occurring, given that another event has already occurred. For example, if an earthquake occurs that results in a landslide, then the probability of the landslide being caused by an earthquake is a conditional probability.

Probabilities of events occurring can be defined as discrete values, or as continuous distributions, as discussed below.

6.4.1 Discrete probabilities

Discrete probabilities can be assigned to events that have either mutually exclusive conditions such as a "stable" slope or an "unstable" slope, or to events with multiple branches such as "no damage", "minor damage" or "major damage". That is, for the mutually exclusive condition, the probability of a stable slope is $P=0.8$ and that of an unstable slope is $P=0.2$ for a total node probability of 1.0. For a multiple event node, the probabilities may be no damage: $P=0.7$, minor damage: $P=0.25$, and major damage: $P=0.05$, again for a total node probability of 1.0.

When these discrete probabilities are combined to model a series of events, it is possible to find the joint (or path) probability of each possible combination of events (see Figure 6.3). For example, an initiating event such as an earthquake may result in a landslide, and the landslide may damage a highway resulting in interruptions to traffic of varying magnitude. The occurrences of these events are expressed as conditional probabilities of mutually exclusive events such that the sum of probabilities at each node is equal to 1.0. That is, if the annual probability of an earthquake of a certain magnitude is $P_{\text{earthquake}}$, then the probability of no earthquake occurring in that year is $[1 - P_{\text{earthquake}}]$. Similarly, the slope along the highway is either stable, or unstable, with mutually exclusive probabilities assigned to each event (P_{stable} and P_{unstable}), and the probabilities of traffic delays are $P_{\text{no delay}}$, $P_{\text{minor delay}}$ and $P_{\text{major delay}}$.

The event tree can then show, for example, the joint probability of a major traffic delay due to a slope being unstable caused by an earthquake as the product:

$$[P_{major\ delay} = (P_{unstable\ slope} \cdot P_{earthquake})].$$

This joint probability is also termed the path probability. Calculation of the path probability of each branch of the tree will allow the relative hazards of the site to be ranked, which is valuable in optimizing mitigation work.

Preparation of event trees requires that reliable estimates, or calculations, of probability be made. As discussed in Section 3.3 – Expert Opinion and Subjective Probabilities, and Section 3.4 – Assessment of Subjective Probabilities, methods of determining probabilities may depend on the stage of the project development. In the early stages of a project when little information is available, estimates of probability will have to be made based on experience and judgement, possibly supplemented by advice from panels of expert who may be more objective than those directly involved in the project. However, as discussed in Sections 3.3 and 3.4, care must be taken to avoid bias and overconfidence in the opinions of the expert panel.

As the project advances and investigation work is carried out, information specific to the project will be collected, and prior knowledge may become available that can be used to update the earlier probability estimates. Examples of prior knowledge are records of transportation systems of slope instability and the effect of these events on operations. Many highway and railway systems keep detailed records of geotechnical conditions which can be used to generate probabilities of events such as the number of slides per year.

6.4.2 Continuous probabilities

Discrete probability values discussed in Section 6.4.1 above are likely to have some uncertainty, unless they are obtained from extensive records of past event occurrences and their consequences. When uncertainty in discrete probabilities is considered to be significant to the results of the event tree analysis, the probabilities could be defined as continuous distributions comprising the most likely value, and estimates of the maximum and minimum values that encompass the expected range of uncertainty. As discussed in Section 1.6.1, an appropriate probability distribution for geotechnical engineering is the Beta distribution where the most likely and the maximum and minimum values are defined and does not extend to infinity as is the case with the normal distribution, for example. It is considered that defining distributions by most likely, maximum and minimum values based on judgement and experience is more reliable than using values for the average and standard deviation when little numeric data for the project is available (see also Section 2.3.3 above on the three-sigma rule to define standard deviation).

The advantage of using continuous distributions, rather than discrete probabilities, is that analysis of the consequences of events will use distributions that more realistically define uncertainty in the parameters. This allows decisions to be made on the appropriate level of mitigation to be used, commensurate with the type of project. For example, for a project where the possible consequences of a failure are severe such as loss of life, then the mitigation work would be selected such that the risk of instability after mitigation is very low. In contrast, where the consequences of failure are low with small financial costs, then the mitigation measures would be limited.

6.4.3 Joint probability

When the occurrence of an event results in another event and the occurrence of both events is uncertain, then the joint probability of the two events can be calculated from the product of the two probabilities. This is shown in Figure 6.3 where the joint probabilities are calculated at the end of each branch. For example, at Branch 3, $[P_{joint}=(P1 \cdot P3)]$ and at Branch 6, $[\mathrm{P}_{\mathrm{joint}}=((1-P1) \cdot (1-P5))]$.

If the two probability values used to calculate a joint probability are continuous probability distributions, then the two distributions can be multiplied together to find the distribution of the joint probability.

For example, Figure 6.4 shows probability distributions for the two events – an earthquake in a specified period (most likely probability=0.2), and slope instability resulting from the earthquake ground motions (most likely probability=0.3). The product of these two distributions is the joint probability that the slope will be unstable if subjected to earthquake ground motions (most likely probability of instability=(0.2·0.3)=0.06). The earthquake probability is shown as a symmetric distribution with a most likely (mode) value of 0.2 and maximum and minimum values of 0.25 and 0.15 respectively. The slope instability probability is shown with a most likely value of 0.3, but is skewed to the right indicative of the high expected likelihood of a slide occurring.

Calculations with probability distributions can be readily carried out by Monte Carlo analysis using software such as @Risk. For example, if two probabilities are defined by Beta distributions, it is possible to add, multiply or divide the distributions to obtain new distributions for the calculated values. Monte Carlo analysis, which involves making multiple calculations of parameter values that are selected at random from the probability distributions, is described in detail in Section 4.5.3, and Figures 4.7 and 4.8 show examples of probability distributions of input and output parameters.

The distribution of the joint probability of a slide caused by the earthquake shows that the most likely probability of the slide occurring is about 6%, with a range from 2% to 10%, and that this distribution is approximately symmetric.

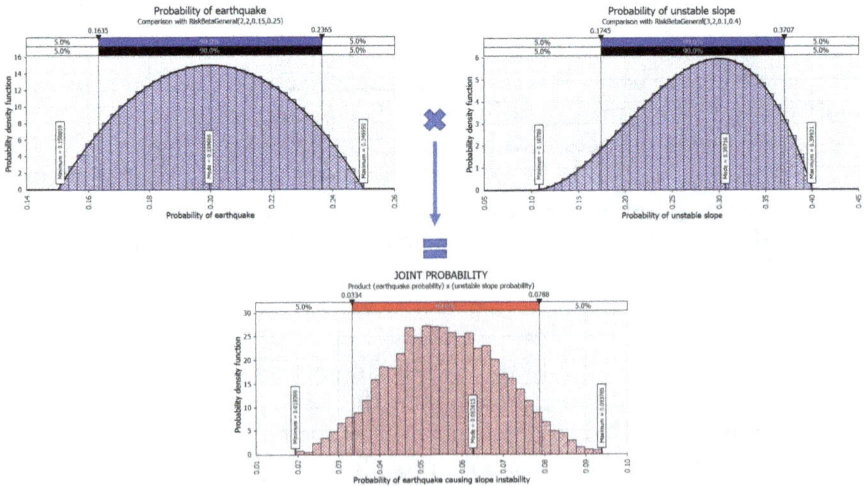

Figure 6.4 Product of probability distributions showing joint probability of an earthquake causing slope instability.

6.4.4 Temporal and spatial probability

A component of many studies of event probabilities is whether persons or property will be within the area of the hazard at the time that the event occurs – temporal probability, or are located within the hazard area – spatial probability (Roberds, 2005).

A common situation for ***temporal probability*** is the difference between a permanently occupied residence that would have a temporal probability of:

$$[P_{\text{house occupied}} = 1.0],$$

and a facility that is only inspected for a limited period each year, and is vacant the remainder of the time so that the temporal probability is less than 1.0. For example, if the duration of the inspection is 8 hours, and is made four times per year for a total of 32 hours, then the annual temporal probability of the facility being occupied during a hazardous event is:

$$[P_{inspection\ occupied} = 32/(365 \cdot 24) = 3.65E-3].$$

If the hazard at this location is slope instability, for which records show that they occur, on average, every five years, then the annual temporal probability of a slide is:

$$[P_{slide} = 0.2].$$

For these conditions, the joint probabilities of persons in the residence being at risk from an event that occurs with an annual temporal probability of 0.2 would be:

$$[P_{temporal/house} = (0.2 \cdot 1.0) = 0.2],$$

and for personnel at the inspection facility:

$$[P_{temporal/inspection} = (0.2 \cdot 3.65E-3) = 7.31E-4].$$

Regarding ***spatial** probability*, consider the situation with the house and the inspection facility discussed in the previous paragraphs, which both have a length of 15 m in the direction parallel to the slope. If the structures are located at the base of a steep slope that is 300 m long and the probability of a slide occurring is equal over the full length of the slope because of the similar geology across the site, then the spatial probability of the structures being impacted by the fall is:

$$\left[P_{spatial/structures} = 15/300 = 0.05\right].$$

The joint probability of both the temporal and spatial probabilities is found from the product of the two probabilities. That is, the joint annual probability of a slide occurring that impacts the house when persons are in the residence is given by:

$$[P_{impact/house\ occupied} = (0.2 \cdot 0.05) = 0.01 = 1E-2],$$

while the joint annual probability of the inspection facility being occupied during a slide is:

$$[p_{impact/inspection\ occupied} = (7.306E-4 \cdot 0.05) = 3.65E-5].$$

Study of temporal and spatial probabilities of hazards allows both situations to be accounted for in a rigorous manner, incorporating all factors that may contribute to the hazard. The results of such a study will often show the relative hazard levels of each structure so that remedial work can be prioritized, in this case for the house rather than the inspection facility.

The acceptability of a calculated probability can be assessed by comparing the calculated annual probability of occurrence with acceptable societal hazards as defined by the [F – N] diagram discussed in Section 5.2 (see Figure 5.3). The [F – N] diagram shows that an annual probability of an accident at the inspection facility of [3.65E-5] is within the low part of the ALARP zone for one or two persons at risk so that the inspections could probably be allowed to proceed without mitigation measures. However, annual probability of impact to the house of [1E-2] would not be acceptable.

6.4.5 Vehicle hazards – railways and highways

A special case of temporal and spatial hazards relates to transportation systems, railways and highways, where operations can be very sensitive to events such as landslides located above and below the track or highway. The consequences of events on transportation systems can include direct costs such as damage to vehicles and injuries to persons, as well as damage to the track and highway, and a wide range of indirect consequences such as interruptions to traffic, repairs to adjacent infrastructure such as utilities, and any legal cases arising from these situations.

Study of temporal and spatial risks on transportation systems needs to consider conditions that include the following (Roberds, 2005):

- **Railways** – trains are much less vulnerable to damage than cars, as shown in the upper image in Figure 6.5 where the locomotive was stopped and derailed by the impact but minimal damage occurred. Trains have the disadvantages that they have long stopping distances and are not able to swerve to avoid hazards. Also, it can happen that the freight or passenger cars continue moving after the locomotive has stopped and come to rest in front of the locomotive. Generally, train traffic will be less frequent and slower than highway traffic, and for a freight train, the most severe hazard is to the

Figure 6.5 Consequences of rock falls on railways and highways; (a) freight train collision with stationary rock; (b) rock fall in flight impacted windshield of moving car (image by N. Boultbee).

front end locomotive with little hazard to the freight cars. However, passenger trains are at risk over their full length and severe consequences can occur if accidents occur at high speed. An example of the precautions that are needed to protect high-speed rail lines is the Shinkansen (Bullet Trains) in Japan that travel at 320 km/hour. The tracks are mostly located in either fully lined tunnels with concrete protection sheds at the portals, or on viaducts such that rock fall and landslide hazards are essentially eliminated. The reliability of these safety measures is demonstrated by the fact that the Shinkansen system has never experienced a fatal accident since operations started in 1964.

- **Highways** – the combined temporal and spatial probabilities of damage to highway vehicles depend on such factors as traffic volume, speed, sight stopping distance (curvature) and the number of traffic lanes. These conditions will influence whether a vehicle will collide with an obstacle on the road, or the much less likely situation that the vehicle will impact a rock fall in flight, as shown in the lower image in Figure 6.5. Another operational condition is that traffic may be stopped on the road because of previous accident or weather conditions, in which case stationary vehicles are generally at higher risk than moving vehicles (Bunce, Cruden, & Morgenstern, 1997).

Details of probabilistic methods of calculating temporal and spatial hazards on highways, together with examples from Italy, Hong Kong, and North Carolina are discussed by Roberds (Roberds, 2005). These discussions include the potential for loss of life (*PLL*) as a result of landslides. For example, in Hong Kong it has been found that the average number of fatalities if a cut slope fails is 0.012, which represents one fatality for each 80 landslides. Also, records have been drawn up for the *PLL* as a result of landslides for various land uses such as buildings – *PLL*=3 to 6 for high occupancy buildings, to *PLL*=0.001 for country parks. The risk management work in Hong Kong shows the value of collecting records over a long period of time that can be used to implement rational mitigation measures.

6.5 EVENT TREE EXAMPLE – FOUNDATION STABILITY

To illustrate the use of event trees to examine the probability of system failure, consider a concrete dam foundation where spillway discharge could develop a scour cavity downstream of the dam resulting in loss of foundation stability (Figure 6.6). Analysis of this situation can be carried out using an event tree to show each component of the foundation, and the

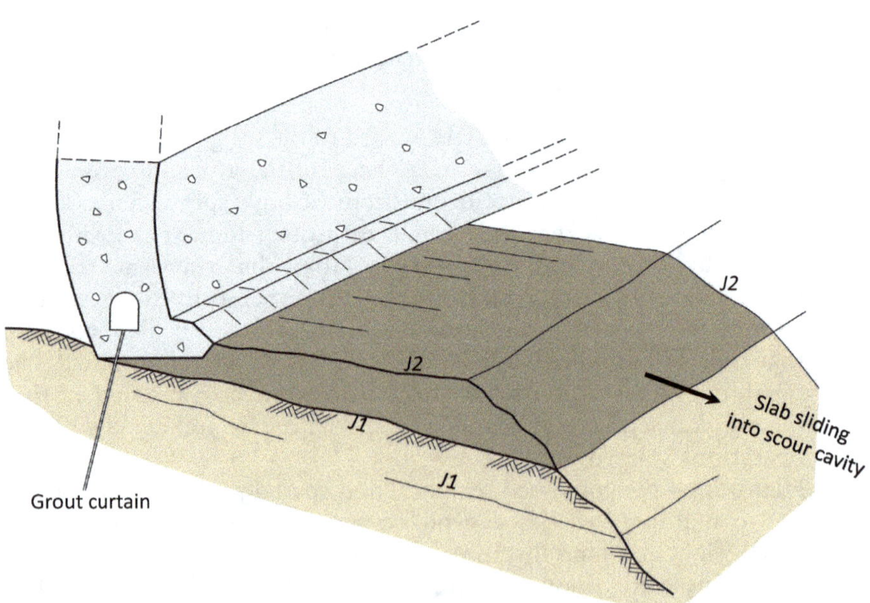

Figure 6.6 Potentially unstable blocks of rock formed in foundation by J1 and J2 joint sets – blocks can slide into scour cavity; grout curtain reduces water pressures in foundation.

possibility of its influence on stability. Examples of scour cavities caused by spillway discharge are Oroville Dam on the Feather River in California (California Department of Water Resources, 2018), and Kariba Dam on the Zambezi River on the border between Zimbabwe and Zambia as discussed in Section 1.5.2. In the case of Oroville, failure of the spillway during flood discharge resulted in scour under the spillway, whereas at Kariba, spillway discharge scoured a cavity in the foundation.

Foundation stability is affected by both external and internal conditions. External conditions are the water force acting on the dam and any earthquake ground acceleration, and internal conditions are the geology of the foundation, rock strength and groundwater pressures. Also, these conditions can be defined according to those that have specific values (epistemic conditions), and conditions that have uncertain values defined by probability distributions (aleatory conditions). For example, known conditions are the specific values for the external water force on the dam, and the mass of the structure, while conditions that have uncertain values are the foundation geology and rock strength, scour depth, rock failure mechanism, groundwater pressure, and occurrence of earthquake ground acceleration. Properties of the foundation rock, and uncertainties in these properties, would be based on mapping and strength testing carried out during investigation, design and construction of the dam. Figure 6.7 shows the event tree for the dam foundation stability analysis comprising three components as follows:

- **Initiation event** – formation of a scour cavity resulting from spillway discharge would be an initiating event, possibly resulting in a system failure of the foundation.
- **Subsequent events** – stability of the foundation depends on whether the joints form blocks of rock, and if the blocks can slide into the scour cavity undermining the dam foundation.

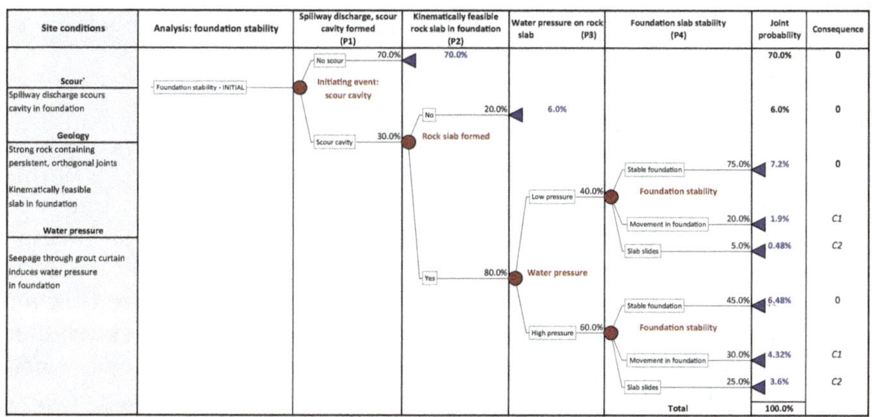

Figure 6.7 Event tree for foundation stability analysis of dam shown in Figure 6.6 for initial conditions (plot generated by Precision Tree, Lumivero Corp.)

- **Consequences** – the consequence of foundation instability could be damage to the dam, as well as destruction downstream of the dam. For simplicity, it can be assumed that three consequences can occur – a stable foundation with no consequence (*0*), movement of a rock block resulting in minor damage to the dam (consequence *C1*), or failure of a portion of the foundation resulting in major damage to the dam (consequence C2).

The event tree shows the following four sets of the conditions that influence rock stability, where uncertainty is defined by their probability of occurrence (*P*) – scour cavity development (*P1*), rock block formation (*P2*), water pressure (*P3*) and stability of foundation slabs (*P4*), as discussed below.

6.5.1 Scour cavity in foundation

At the time of construction, if the river bed and the dam foundation are a continuous, sloping rock surface, it is not feasible for blocks of rock to slide from the foundation. However, if water flow from the operation of the spillway were to scour a cavity in the foundation just downstream of the dam, then blocks of rock that "daylight" in the face of the cavity could become unstable – kinematically feasible (Figure 6.6) (Annandale, 1995); (Annandale, Abt, Ruff, & Whittier, 1996). The potential for rock scour depends on both geological factors – rock strength, discontinuity orientation, length and spacing, and on the performance of the spillway to pass flood flows without exposing the rock to scour. All these conditions have some degree of uncertainty, and a probability distribution (*P1*) could be established to define the probability of a scour cavity developing.

Probability of scour developing – study of the potential for scour potential in a dam foundation using event trees is shown in Figure 6.7, initial conditions and Figure 6.9, upgraded conditions. For initial conditions, the probability of a cavity being scoured in the foundation is estimated to be *P1*=30%. However, if the spillway were to be upgraded, the probability of a scour cavity developing can be reduced to *P1*=10%.

6.5.2 Rock blocks formed in foundation

Geology of the foundation will be established from site investigations during the feasibility stage of the project, foundation models used in design, and then mapping of the exposed rock during construction. The site information will consist of the shear strength of the discontinuities and the structural geology, with stability being of concern if the discontinuities are oriented to form three-dimensional blocks that are kinetically feasible and can slide in the downstream direction out of the foundation. Also of concern is persistence of the discontinuities, and if the persistence is sufficient to create a block large enough to influence the stability of the foundation (Figure 6.8).

Despite the site investigation work, uncertainty will remain of rock properties at depth in the foundation that cannot be directly observed, and because rock can weather and its strength deteriorate with time.

Probability of unstable block formation – probability of kinematically feasible blocks in the foundation is estimated to be $P2=80\%$ based on the known properties of *J1* and *J2* joints mapped during the site investigation, where *J1* joints dip downstream. The probability of unstable block formation is the same before and after upgrading the spillway because the geology is unchanged.

6.5.3 Water pressure on rock wedge

A high water pressure gradient will exist in the foundation between the full reservoir head at the upstream side, and the tailwater pressure downstream of the dam. Water pressure in the foundation, as well as seepage through the foundation, can be controlled by installation of a grout curtain comprising rows of holes drilled in the foundation rock into which cement or chemical grout is pumped under pressure. An essential feature of grout curtains is that the grout penetrates and seals fractures in the rock in which water can flow, but it is unlikely that complete sealing can be achieved. Piezometers can measure water pressures upstream and downstream of the grout curtain but only at discrete points that may not be representative of the entire curtain (Cedergren, 1989). Therefore, the grout curtain may be incomplete, and uncertainty will exist regarding the water pressure acting on blocks in the foundation.

Probability of water pressure acting on block in foundation – based on the likely effectiveness of the grout curtain and drainage measures, it is estimated that the probability of a high water pressure in the foundation slabs for initial conditions is $P3=60\%$. However, improvements could be made to the grout curtain by additional grouting from the gallery in the base of the dam such that the probability of high water pressure in the foundation is reduced to $P3=40\%$ (Figure 6.9).

6.5.4 Wedge stability analysis

Application of event tree analysis to examine the dam foundation integrity requires that stability analyses be carried out of blocks of rock in the foundation to determine how site conditions such as geological structure, rock shear strength and groundwater pressures influence stability. Instability is possible if discontinuities are oriented to form three-dimensional blocks that can slide in the downstream direction out of the foundation into the scour cavity, and if the persistence of the discontinuities are sufficient to create blocks that are large enough to influence stability of the foundation. Figure 6.8 shows a potentially unstable (kinetically feasible) slab of rock that can slide on *J1* joints that dip downstream, with steep *J2* joints oriented parallel to the valley forming release surfaces on the sides of the slab that provide no shear resistance. The persistence of the *J1* and *J2* joints are sufficient for the slab to extend from the face of the scour cavity to beneath the dam so the stability of the dam would be in jeopardy if the slab were to displace.

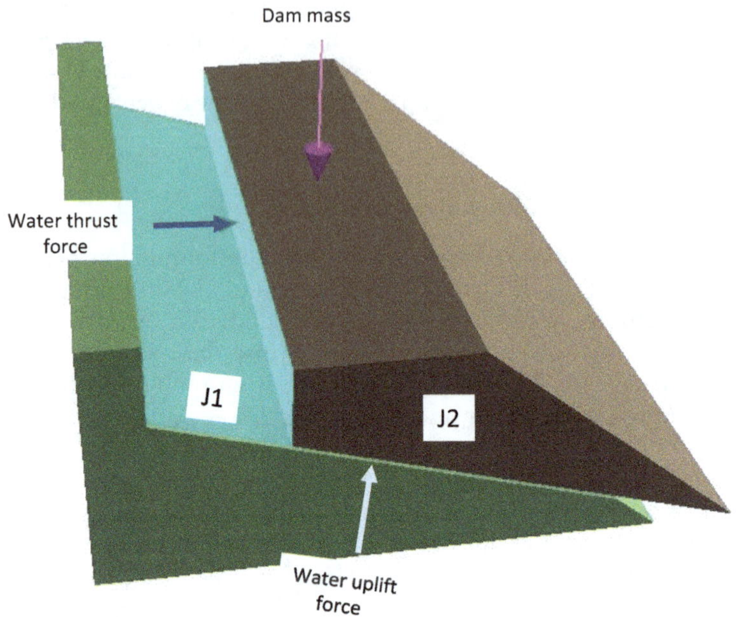

Figure 6.8 Stability analysis of rock slab in dam foundation using software ROCPLANE (RocScience Inc.).

Simplified stability of the foundation slabs can be studied using limit equilibrium analysis, for which suitable software is ROCPLANE (RocScience Inc., 2021). Figure 6.8 shows a model of a planar slab in the foundation formed by the *J1* joint set that dips downstream at a dip angle of 18°. Because most of the parameters in the planar stability analysis have uncertain values, the analysis can incorporate probability distributions for these parameters to calculate both the probability of failure and the factor of safety (see Chapter 2 above).

The ROCPLANE analysis will show the influence on stability of groundwater pressures in the foundation. Possible results of the stability analysis are that the slabs are stable (Factor of Safety > 1), that the slabs are marginally stable (Factor of Safety ≈ 1) resulting in possible movement of the foundation, or that the Factor of Safety < 1.0 in which case the foundation slabs may slide.

These results can be used as input values in the event tree analysis (see Figure 6.7).

6.5.5 Results of event tree analysis

Initial spillway condition stability analysis: the event tree in Figure 6.7 shows the structure of the initial stability study, and the estimated probability of occurrence for each of the four conditions – scour cavity, rock blocks, water pressure, wedge stability - influencing stability as discussed in

Sections 6.5.1 to 6.5.4 above. Information on probable stability conditions can be obtained from the event tree as follows, where the joint probabilities are shown in the second column from the right in the event trees:

- The estimated probability that the foundation will be resistant to scour and be stable is [$P1_{\text{no scour}}=0.70$]. If scour does occur [$P1_{\text{scour}}=0.30$], the probability that no potentially unstable slabs will be formed is [$P2_{\text{no slabs}}=0.20$] based on the geological conditions in the foundation of joints dipping downstream.

  Joint probability that a scour cavity will develop but the foundation will be stable is:

  $$[P_{stable} = 0.30 \cdot 0.20 = 6\%].$$

- If a scour cavity develops, the probability that potentially unstable slabs will be formed in the foundation is:

  [$P2_{\text{slabs}}=0.80$], based on the foundation geology with persistent *J1* joints dipping downstream.

  Also, the probability that low water pressure will develop in the foundation is:

  [$P3_{\text{low pressure}}=0.40$] based on the grout curtain condition.

  Stability analyses of the foundation show that, for these conditions, the probability that the foundation will be stable is:

  $$\left[P4_{stable} = 0.75\right],$$

  and the probability that movement may occur is:

  $$\left[P4_{movement} = 0.20\right]$$

  and that the slabs will slide is:

  $$\left[P4_{slide} = 0.05\right]$$

  The "Joint probability" column in Figure 6.7 shows the calculated joint probabilities for these stability conditions. Joint probability of an unstable foundation where the rock blocks slide is along the path:

  $$[P_{sliding\ slab} = 0.3 \cdot 0.8 \cdot 0.4 \cdot 0.05 = 0.48\%].$$

- The other possible stability condition is the probability that high water pressure develops in the foundation [$P4_{\text{high pressure}}=0.60$] with the result that the foundation slabs will be less stable than for the low-pressure condition with a probability that slabs will slide:

$\left[P4_{slide} = 0.25\right]$.

Joint probability of slabs sliding for the high water pressure condition is along the path:

$$[P_{sliding\ slab} = 0.3 \cdot 0.8 \cdot 0.6 \cdot 0.25 = 3.6\%]$$

- The "Joint probability" column in Figure 6.7 shows that the three foundation stability conditions are "stable", "movement" and "sliding", with corresponding consequences of "0", "C1" and "C2". The total probability of occurrence for each of these conditions can be found by adding the joint probabilities.

  Total probability that the foundation will be stable is:

$$\left[P_{stable} = (0.7 + 0.06 + 0.072 + 0.0648) = 89.7\%\right].$$

- Similarly, the total probabilities that foundation slabs will "move" or "slide" can be found by adding the corresponding joint probabilities.

  For the two conditions where the foundation blocks will "move" is:

$$\left[P_{move} = (0.019 + 0.0432) = 6.22\%\right]$$

For the two conditions where the foundation blocks will "slide" is:

$$\left[P_{slide} = (0.0048 + 0.036) = 4.1\%\right]$$

These three mutually exclusive events have a total probability of occurrence of:

$$[P_{total} = (89.7 + 6.2 + 4.1 = 1.0]$$

Upgraded spillway condition stability analysis – another use of event tree analysis is to assess how changes to the probability of occurrence of events can change the probability of the foundation being stable (Figure 6.9). For example, the spillway can be upgraded as follows:

- Strengthen resistance of the spillway to scour so that the probability that a scour cavity will develop will be reduced from:

$$\left[P1_{scour\ (initial)} = 30\%\right]$$

to

$$\left[P1_{scour\ (upgrade)} = 10\%\right]$$

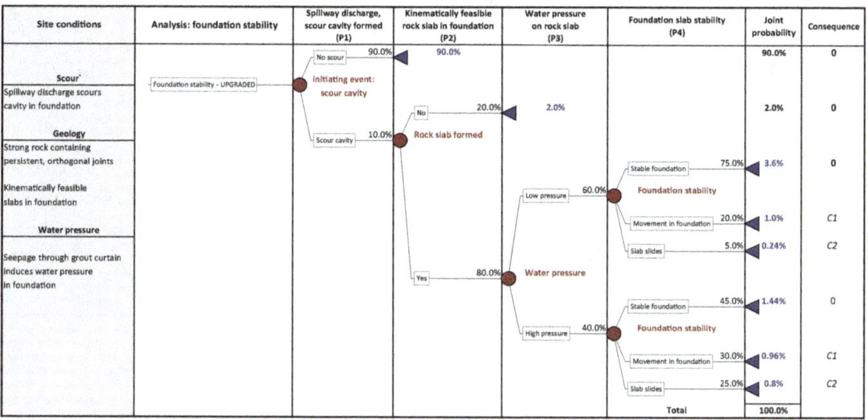

Figure 6.9 Event tree for spillway scour after upgrades to increase scour resistance, and to improve grout curtain with reduction in foundation water pressure (plot generated by Precision Tree, Lumivero Corp.).

- Improve grout curtain with an additional row of grout holes so that the probability of high water pressures in the foundation is reduced from:

  $$\left[P4_{high\ pressure\ (initial)} = 60\%\right]$$

  to

  $$\left[P4_{high\ pressure\ (upgrade)} = 40\%\right].$$

When these new probability values are input in the event tree, the probability of the foundation being stable is increased from:

$$\left[P_{stable\ (initial)} = (0.7 + 0.06 + 0.072 + 0.0648) = 89\%\right]$$

to

$$\left[P_{stable\ (upgrade)} 0.9 + 0.02 + 0.036 + 0.0144 = 97\%\right].$$

Probability that the foundation will slide is reduced from:

$$\left[P_{slide\ (initial)} = (0.0048 + 0.036) = 4.1\%\right]$$

to

$$\left[P_{slide\ (upgrade)} = (0.0024 + 0.008) = 1.0\%\right].$$

These analyses will provide guidance on the most effective remedial work to be carried out. That is, the probability of the foundation being stable is increased from 89% to 97% by improving scour resistance and reducing the water pressure. The cost of the improvement work would need to be compared to the degree of increased stability.

The event tree structure can be imported into decision analysis in which probability distributions for the costs of remedial work and the consequences of events can be analyzed to determine an appropriate remedial action. Decision analysis is discussed in Chapter 7.

Chapter 7

Decision analysis

Analysis of risk for geotechnical projects will usually show uncertainties in the likelihood of event occurrences, and/or the consequence(s) of the event, where the risk is the product of the occurrence and consequence probabilities. As discussed in Chapter 5, the calculated risk can be compared to acceptable levels of risk based on societal norms or on corporate risk standards. If the risk is deemed to be unacceptable, then a decision has to be made on how to implement remedial measures to reduce the risk to an acceptable level. Risk can be reduced by reducing the likelihood of occurrence and/or reducing the consequence. If the risk cannot be reduced to an acceptable level, because, for example, the cost of the required mitigation cannot be justified, the project may need to be abandoned.

This chapter discusses decision analysis and how it can be applied in geotechnical engineering to analyze alternative actions – Task 6 in Figure 7.1. For example, if the risk of slope instability is unacceptable because people living in houses below the slope are in danger, then a decision will have to be made to reduce the risk. The likelihood of a slide could be reduced through different means such as installing drains to reduce the groundwater level in the slope, unloading the crest of the slope, and/or by installing tie-backs to reinforce the slope. The consequence of a slide could be reduced by evacuating the houses in the slide runout path or installing a diversion structure to prevent slide material from reaching the houses. For these circumstances, drainage is likely to be the least expensive action, but the reliability of drainage is less than that of installing tie-backs or evacuating the houses. A rational method is required to select the action that achieves an acceptable level of risk.

Decision Analysis is a rational procedure for analyzing alternative mitigation measures in which both the uncertainties in event occurrence and of the consequences of these events can be combined to determine the "Expected Value" (*EV*) of the event. The expected value is the product of the probability of occurrence and its consequence. For example, a very low probability event that has a potentially high consequence will have a low expected value. By adding the calculated expected values of consequences,

DOI: 10.1201/9781003271864-7

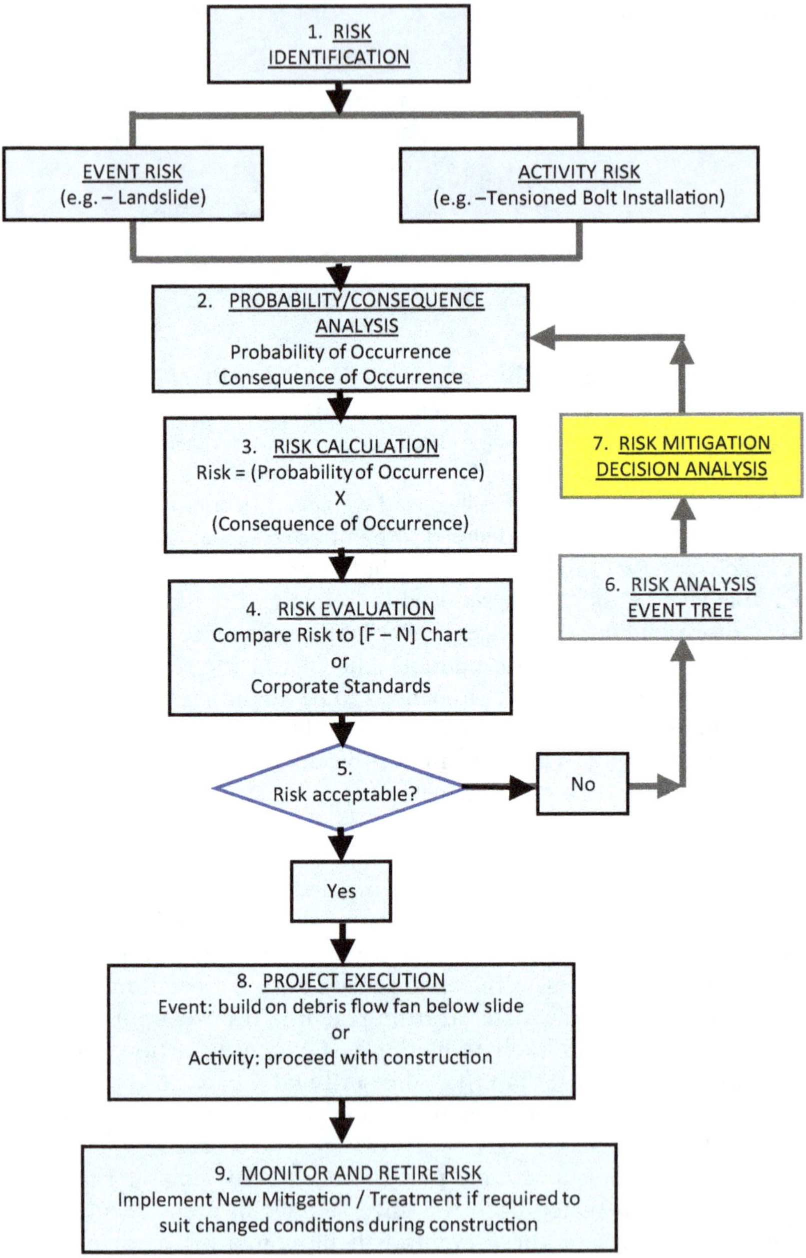

1. RISK IDENTIFICATION

EVENT RISK (e.g. – Landslide)

ACTIVITY RISK (e.g. –Tensioned Bolt Installation)

2. PROBABILITY/CONSEQUENCE ANALYSIS
Probability of Occurrence
Consequence of Occurrence

3. RISK CALCULATION
Risk = (Probability of Occurrence)
X
(Consequence of Occurrence)

4. RISK EVALUATION
Compare Risk to [F – N] Chart
or
Corporate Standards

5. Risk acceptable?

No

6. RISK ANALYSIS EVENT TREE

7. RISK MITIGATION DECISION ANALYSIS

Yes

8. PROJECT EXECUTION
Event: build on debris flow fan below slide
or
Activity: proceed with construction

9. MONITOR AND RETIRE RISK
Implement New Mitigation / Treatment if required to suit changed conditions during construction

Figure 7.1 Flow chart of risk management tasks - Task 7, decision analysis, select optimum mitigation method.

and the costs for mitigation, for each mitigation option, it is possible to identify the optimum mitigation strategy.

That is, the expected value of a decision is:

$$EV_{decision} = \Sigma((probability\ of\ occurrence) \cdot (consequence\ cost) + (mitigation\ cost))$$

Once an event tree has been developed for a project to show the relationship between initiating events, such as earthquakes and possible subsequent events such as landslides, it is possible to use the tree to carry out decision analysis in which options for remedial action are analyzed. The objective of the analysis is to identify the remedial action and mitigation measure with the optimum Expected Value, which may be the lowest cost consequence in the case of a hazard such as a landslide, or the minimal cost overrun in the case of a construction project.

7.1 DECISION TREES: DECISION, CHANCE AND END NODES

Decision trees, which have a similar structure to event trees (see Chapter 6), are made up of three types of nodes as follows (see Figure 7.2):

decision nodes (squares). Branches originate at decision points defining the possible actions that may be taken, with the branches connecting to chance nodes that show the possible events resulting from each decision. Each branch from a decision point shows the cost of taking that action such as stabilization work or an investigation program.

chance nodes (circles) from which branches (lines) originate, with each branch showing the probability that an event will occur.

end nodes (triangles) at the termination of each branch where consequences of events and actions are documented.

The objective of the decision analysis is to find the most favourable action for the project, which is shown on the decision tree as "TRUE" for the most favourable action, with "FALSE" showing action(s) that are less than favourable. For geotechnical projects, both the cost of taking an action such as stabilizing a slope, and the consequences of taking the action, such as an interruption to traffic, will be negative values. Therefore, the most favourable (TRUE) decision is that with the lowest negative number.

Decision nodes are located at the start of the decision tree and can be at intermediate points where new decisions are required based on, for example, information collected after the first decision node.

Figure 7.2 illustrates a decision tree to evaluate whether a potentially unstable slope should be stabilized at a cost of (–$200,000) in order to

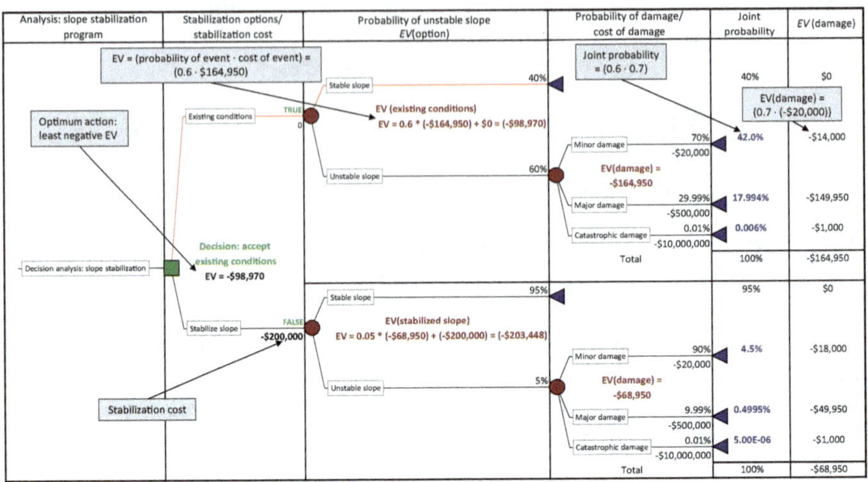

	Joint probability	EV(damage)
Stable slope 40%	40%	$0
Minor damage 70% -$20,000	42.0%	-$14,000
Major damage 29.99% -$500,000	17.994%	-$149,950
Catastrophic damage 0.01% -$10,000,000	0.006%	-$1,000
Total	100%	-$164,950
Stable slope 95%	95%	$0
Minor damage 90% -$20,000	4.5%	-$18,000
Major damage 9.99% -$500,000	0.4995%	-$49,950
Catastrophic damage 0.01% -$10,000,000	5.00E-06	-$1,000
Total	100%	-$68,950

Figure 7.2 Structure of typical decision tree showing probabilities of events, consequences, remediation cost, expected values and optimum decision.

reduce the probability of slope failure. If slope failure were to occur, the possible consequences of failure are quantified in three categories as:

a. minor damage costing (–$20,000);
b. major damage costing (–$500,000);
c. or catastrophic damage costing (–$10 million).

The tree is annotated to identify the main components of the analysis, the features of which are discussed below.

Decision trees discussed in this chapter have the same structure and properties as event trees discussed in Chapter 6 and illustrated in Figure 6.3.

7.1.1 Mutually exclusive events

Several mutually exclusive events may occur at each chance node. That is, a slope may be stable with a probability of [P_{stable}=x], or unstable with a probability of [P_{unstable}=(1 – x)]. The sum of the probabilities at this chance point is [$P_{\text{stable}}+P_{unstable}$=1.0] showing that only these two events are possible. The estimated probabilities of occurrence of three levels of damage resulting from slope instability are also shown in Figure 7.2. That is, for existing conditions, the probabilities of damage occurring in the event of a slide are:

a. minor damage with probability of 70%;
b. major damage with probability of 29.99%;
c. or catastrophic damage with probability of 0.01%.

These are mutually exclusive conditions with a total probability of 1.0.

7.1.2 Joint ("path") probability

In Figure 7.2, a joint (or "path") probability is the resultant probability of the events along a branch from the chance node to the end node. For example, where the probability of an unstable slope is 60% and the probability of minor damage, if a slide occurs, is 70%, then the joint probability of the slide causing minor damage is:

$$[P_{minor\ damage} = (0.6 \cdot 0.7) = 42\%].$$

Similarly, the joint probability of major damage is:

$$[P_{major\ damage} = (0.6 \cdot 0.2999) = 17.994\%],$$

and the joint probability of catastrophic damage is:

$$[P_{catastrophic} = (0.60 \cdot 0.0001 = 006\%].$$

Because the probability that the slope will be stable is:

$$\left[P_{stable} = 40\%\right]$$

the total probability of events resulting from the existing condition node is:

$$\left[P_{total} = 42\% + 17.994\% + 0.006\% + 40\% = 100\%\right]$$

This addition to 100% of event probabilities for each decision provides a useful check that the probability calculations are correct.

Information that is provided by the joint probabilities on the decision tree is the comparative likelihood of events occurring, such as before and after stabilization:

- **Existing conditions** – probability that the slope will be unstable is:

  $$\left[P_{unstable} = 60\%\right],$$

  and the probability of an event causing minor damage is:

  $$[P_{minor\ damage} = (0.6 \cdot 0.7) = 42\%]$$

- **After stabilization** - probability that the slope will be unstable is:

  $$\left[P_{unstable} = 5\%\right],$$

  and the probability of an event causing minor damage is:

  $$[P_{minor\ damage} = (0.05 \cdot 0.9) = 4.5\%]$$

These values for the joint probability of minor damage show that the stabilization work may reduce the probability of minor damage by an order of magnitude (42% to 4.5%), which is indicative that the stabilization work is effective with a low level of uncertainty in its performance.

7.1.3 Consequences of events

Each branch of the decision tree will have a consequence of event occurrence. Depending on the data available on consequences, it is possible to assign a consequence to each branch, or only to the terminations of the branches to show the total consequence of the sequence of events. Consequences can be either positive such as savings of cost or time, or negative such as injuries, lives lost, costs incurred or schedule delays.

For the application of decision analysis, it is usual to express the costs of remedial actions, and of the consequences, as monetary values. This allows the costs of remediation and consequences to be added.

In Figure 7.2, consequences are shown for three possible events resulting from a landslide:

a. Minor damage to a house of (–$20,000);
b. Major damage such as destruction of a house costing (–$500,000); or
c. Catastrophic damage such as destruction of multiple buildings with casualties to inhabitants costing (–$10 million).

It is assumed that the damage costs are the same both before and after stabilization work, but their probability of occurrence is decreased by installation of remedial work.

Consequence tree – consequences may be single values, or a consequence tree may be developed that shows a sequence of consequences resulting from an event. An example of a consequence tree is repair work required if scour of a highway bridge foundation causes settlement to occur. Types of remedial work may be both direct and indirect - direct work on the bridge comprises the improvement of the foundation and adjustments to the bridge deck, while indirect effects comprise interruptions to traffic during the site work, and possibly repairs to vehicles that were damaged when travelling over the bridge after settlement occurred (Figure 7.3).

Assignment of accurate consequences to events is important if the event tree is part of a study to select optimum remedial measures. For example, the cost of remedial measures such as inspection and maintenance of the bridge foundation and deck should be consistent with the traffic volume and loads. That is, it is expected that remedial expenses would be significantly greater for a major highway bridge than for a bridge on a lightly travelled industrial road.

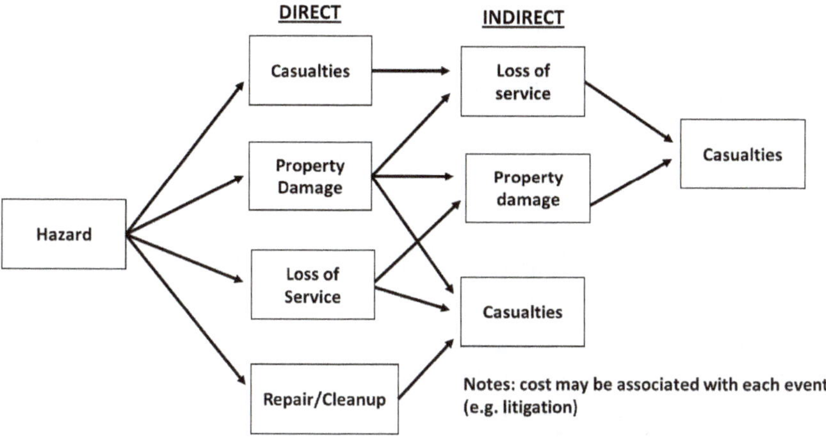

Figure 7.3 Examples of direct and indirect consequences of a hazard occurrence (Roberds, 2005).

It is also possible to express the consequences as discrete values, or as probability distributions that account for the uncertainty in the costs of the consequences, as discussed in Section 7.3 below.

7.1.4 Expected value (EV) of events

The probability of an event occurrence and its consequence can be combined by taking the product of the two values, which is termed the Expected Value (EV) defined as:

$$EV = \left(probability\right) \cdot \left(consequence\right) \tag{7.1}$$

For example:

Expected Value of landslide ($EV_{\text{landslide}}$)

= (probability of landside occurrence) · (consequence of landslide)

A key feature of expected value calculations is that they account for the combined influence of both the probability and consequence of events. That is, a low probability but high consequence event may have a low EV value and be of less importance than higher probability, but lower consequence events. For example, as shown in Figure 7.2, it is considered that a catastrophic event has a very low probability of occurrence of [P=0.01%], but the consequence damage would be significant at an estimated cost of (–$10 million). The EV for the catastrophic event is:

$$[EV_{catastrophe} = (1E-4 \cdot (-\$10{,}000{,}000)) = (-\$1{,}000)].$$

In comparison, the estimated probability of a minor slide occurring is 70%, for which the damage would be (–$20,000), and the EV is:

$$[EV_{minor\ slide} = (0.7 \cdot (-\$20{,}000)) = -\$14{,}000].$$

This result demonstrates that minor slides have a higher expected value, and risk, than a catastrophic event.

7.1.5 Calculation of Expected Values using joint probabilities

Joint probabilities can also be used to calculate the expected value of each decision point using the following relationship:

$EV = [\Sigma((\text{joint probability}) \cdot (\text{consequences along branch}))]$

For example, in Figure 7.2 the Expected Value of existing conditions can be calculated from the joint probabilities and the costs of minor, major, and catastrophic damage as follows:

$$[EV_{existing\ conditions} = 0.42 \cdot (-\$20{,}000) + 0.17994 \cdot (-\$500{,}000) + 6E-4 \cdot (-\$1E6) = (-\$98{,}970)]$$

Similar joint probability calculations are shown in Figures 7.5, 7.6 and 7.8 to 7.11.

7.1.6 Expected value of decisions

The decision tree in Figure 7.2 examines the decision to either accept existing conditions, or to stabilize the slide, taking into account the uncertainties in site conditions . Selection of the optimum course of action involves examining the Expected Values of each option, together with the cost of the stabilization work, to find which option has the least expected cost. This data is summarized below.

- **Existing conditions** – the sum of the expected values of three possible types of instability is:

$$[(\Sigma EV_{damage} = (-\$14{,}000) + (-\$149{,}950) + (-\$1{,}000) = (-\$164{,}950)$$

If the estimated probability of 60% that a slide will occur is considered, then the expected value of damage due to slope instability is:

$$[EV_{instability} = 0.6 \cdot (-\ \$164{,}950) = (-\$98{,}970)].$$

Because no costs are incurred with acceptance of existing conditions, (–$98,970) is the total EV for this option.

- **Stabilize slope** – when stabilization measures have been implemented, the probability of instability occurring is reduced from 60% for existing conditions, to 5%, and the corresponding expected value of damage due to slope instability is:

$$[EV_{instability} = 0.05 \cdot (-\$68,950) = (-\$3,448)].$$

To this expected value of instability is added the cost of the stabilization work of (–\$200,000) for a total expected value for the stabilized slope of:

$$[EV_{stabilized\ slope} = (-\$3,448) + \left(-\$200,000\right) = \left(-\$203,448\right).$$

These two calculated expected values indicate that stabilization work costing (–\$200,000) may not be justified on economic means alone because:

$$\left[EV_{existing} = -\$98,970\right] \text{ is less than } [EV_{stabilized\ slope} = -\$203,448].$$

Analysis of the decision tree shows that if the stabilization work could be reduced to (–\$80,000), then the total EV for the stabilization option is less than the total EV for existing conditions. However, this result is offset by the situation that the reduced stabilization work may be less effective such that the probability of instability of the stabilized slope is greater than 5%. The decision tree can help to find the optimum combination of stabilization work and reliability of the stabilized slope.

Section 7.3 presents a similar decision tree for slope stabilization options, and Section 7.7 presents four decision tree case studies: debris flow protection, slope stabilization for a highway, drilling probe holes in a tunnel and installing a test anchor in a dam.

7.2 VALUE OF LIFE

An important issue regarding consequences when lives may be lost, is the cost of human life, expressed as the value of life (VOL), or the value of statistical life (VOSL). An explanation of these terms is provided below.

Monetary value is not put on individual lives when analyzing public safety. Rather, when conducting cost-benefit analyses for actions such as public safety and environmental protection, government agencies and insurance companies use estimates of how much people are willing to pay for small reductions in their risk of dying from adverse health conditions. Examples of causes for adverse health are accidents or environmental pollution. In the scientific literature, estimates of willingness to pay for small reductions in mortality risks are often referred to as the "value of a statistical life". This is because these values are typically reported in units that match the aggregate monetary amount that a large group of people would be willing to pay for a reduction in their individual risks of dying in a year, such that one fewer death would be expected among that group during the year on average. For example, suppose each person in a sample of 100,000 people was asked how much he or she would be willing to pay for a reduction in their individual risk of dying of 1 in 100,000, or 0.001% over the next year. Because this reduction in risk would mean that one fewer death would be expected among the sample of 100,000 people over the next year on average, this is described as "one statistical life saved". Suppose the average response to the hypothetical question was $100, then the total dollar amount that the group would be willing to pay to save one statistical life in a year would be ($100 per person)×(100,000 people)=$10 million. That is, the "VOSL" is a measure of the populations' willingness to pay for risk reduction and the marginal cost of enhancing public safety – it is not an estimate of how much money any single individual would be willing to pay to prevent the certain death of any particular person.

VOSLs are difficult to define because they depend on their application, and they vary from country to country. In the United States, the Department of Transportation and the Department of Environmental Protection use a VOSL value of about $US9 million USD. For reference, this amount can be compared to actual payments made to families as a result of an accident. For example, after the two crashes of the Boeing 737 MAX passenger airliners in 2018 and 2019 in which 346 people lost their lives, the total cost of the accidents was estimated at $$US18 billion USD, or $US52 million USD per life lost. This amount contrasts with the payout made to the families of the deceased of $145,000 USD per person (Ale, Hartford, & Slater, 2021).

Application of decision analysis based strictly on study of cost-benefits may benefit from a comprehensive examination of probabilities of both the hazards and consequences of events. For example, the Ford Pinto car produced in the early 1970s suffered from a poor safety record that was attributed to cost cutting in manufacturing related to the protection of the fuel tank in rear-end collisions (Shaw & Barry, 2013). In geotechnical engineering, similar situations may arise for projects in the ALARP zone of the [F – N] diagram (Figure 1.3), and how much money should be spent to reduce the probability of lives being lost in order for the risk to be "acceptable".

7.3 EXAMPLE OF DECISION ANALYSIS

Figure 7.4 shows a house located below a marginally stable slope where the house is at risk from severe damage, and the occupants are at risk from injury or death, in the event of a slide. Observation of existing conditions of minor tension cracks at the crest and down-slope bent ("pistol grip") trees indicate that slope movement has occurred, and that the slope may be marginally stable. Furthermore, the presence of groundwater in the slope is probably contributing to instability.

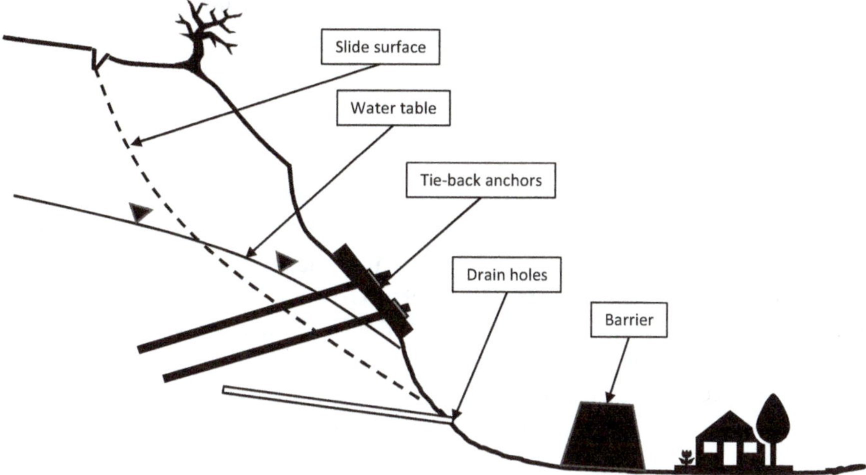

Figure 7.4 Slope stabilization options to protect house from landslide risk.

Possible measures that could be taken to protect the house and its occupants over the long term are to improve stability by lowering the water table by installing drains, or by reinforcing the slope with tie-back anchors. Alternatively, a barrier could be constructed at the base of the slope to protect the house in the event of a slide, with no stabilization work on the slope. Evaluation of these options, in comparison with existing conditions, can be carried out using decision analysis in which the probabilities of a slide occurrence can be combined with the costs of both the remedial measures, and that of damage and injury, to find the optimum course of action. The remediation costs can be compared to the expected cost of a slide where it is estimated that the average consequence of a landslide is (–$200,000).

The probability of a landslide in the future will depend on the method and extent of the remediation measures. That is, slope stability analyses can be carried out using probabilistic methods to calculate the probabilities of slope failure for drilling drain holes, and for installing tie-back anchors. Methods of calculating the probability of slope failure are discussed in Section 2.3 above; Section 2.4 includes a discussion on the meaning of the term "probability of failure" (*PF*) such that the calculated *PF* has no time associated with it, and is just a method of comparing the relative magnitudes and distributions of the resisting and driving forces acting in the slope. Therefore, the *PF* is a convenient method of comparing existing slope stability with stability after drainage, and stability after installing tie-back anchors.

For existing conditions, stability analysis taking into account the weak, fractured rock forming the slope, the presence of groundwater and the evidence of recent movement, shows that the probability of slope failure is about 50%.

The following is a discussion on the two mitigation options to maintain a stable slope, and the calculated probabilities of slope failure for each option.

Option 1, drainage–drill a series of horizontal drain holes to lower the water table in the slope at a cost of (–\$10,000). Drainage of slopes is a well-established method of improving slope stability, but for these conditions has the shortcomings that drains could become plugged with time, and minor slope movement could damage the drains . Additionally, it is difficult to quantify the effectiveness of drains and to guarantee that the water table will be maintained below a specified level, particularly in the event of heavy rainfall.

Probability of slope failure, *PF* – probabilistic stability analyses, which take into account uncertainties in the effectiveness of the drainage system, show that the *PF* after installing drain holes is 30%, and the probability of the slope being stable is 70%.

Option 2, tie-backs–install a pattern of high-capacity tie-back anchors in the lower part of the slope at a cost of (–\$50,000). Support provided by the tie-backs could be verified by the strength of the steel used for the anchors, and the magnitude of the anchoring force as shown by carrying out load-elongation tests for all anchors. The major benefits of the tie-back anchor solution are that the support provided, and the stability condition, can be verified, and that this condition can be maintained over the long term.

Probability of slope failure, *PF*–probabilistic stability analyses, which take into account the reliability of tie-backs compared to drain holes, show that the *PF* after installing tie-back anchors is 5%, and the probability of the slope being stable is 95%.

Figure 7.5 shows the decision tree that examines the cost-benefit of the two mitigation options compared to existing conditions where the estimated cost of the house being damaged by a slide is (–\$200,000). Features of the decision tree in Figure 7.5 are as follows:

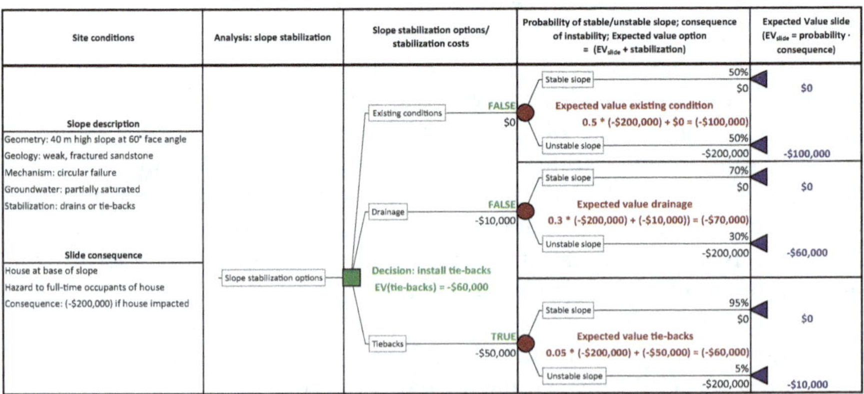

Figure 7.5 Decision tree showing analysis of two slope stabilization options – drainage and tie-back installation for slope shown in Figure 7.4 (PrecisionTree, Lumivero Corp).

a. The square indicates a decision point for the three options showing the cost of implementing the decision, circles are chance points showing the probabilities of a stable and unstable slope; triangles are the termination of each branch where the probability of a stable/unstable slope, and the consequences of slope instability are listed.
b. For each of the three stabilization options and their cost of implementation (negative values), the chance node shows that the slope can be either stable or unstable.
c. The branches on the tree show the calculated probabilities that the slope will be stable for the existing slope, for the drained and for the supported slope. For the existing condition, analysis shows that the slope is marginally stable, and that the probability of instability is 50%. Drainage of the slope should improve stability, but because drainage has limited reliability, it is found that the probability of instability of the drained slope is as high as 30%. Installation of tie-back anchors is a more reliable method of improving stability and the analysis shows that the probability of instability for this option is 5%.
d. For a stable slope, the consequence is zero for each of the three options, but the stabilization costs for drainage (–\$10,000) and tie-back anchor installation (–\$50,000) are still incurred.

Table 7.1 summarizes the probabilities and consequences of instability, the Expected Value (*EV*) of each option and the most favourable option (TRUE).

These results show that the optimal solution is to install tie-back anchors with an *EV* of (–\$60,000) and significantly reduce the probability of slope

Table 7.1 Summary of consequences, remediation work and expected values for slope stabilization options

	Existing conditions	*Drainage using drain holes*	*Installation of tie-backs*
Cost of remedial work	\$0	(–\$10,000)	(–\$50,000)
Consequences of slope instability	(–\$200,000)	(–\$200,000)	(–\$200,000)
Probability of slope instability	50%	30%	5%
Expected value of slope instability (EV_{slide})	$EV_{\text{existing conditions}} = 0.5 \cdot$ (–\$200,000) = (–\$100,000)	$EV_{\text{slide drainage}} = 0.3 \cdot$ (–\$200,000) = (–\$60,000)	$EV_{\text{slide tie-backs}} = 0.05 \cdot$ (–\$200,000) = (–\$10,000)
Expected value of stabilized slope (EV_{slope})	EV_{slope} = (–\$100,000) + (\$0) = (–\$100,000)	EV_{slope} = (–\$60,000) + (–\$10,000) = (–\$70,000)	EV_{slope} = (–\$10,000) + (–\$50,000) = (–\$60,000)
Decision	FALSE	FALSE	TRUE

instability. This result is shown in Figure 7.5 as "TRUE", meaning this is the lowest negative number.

The decision analysis result shown in Figure 7.5 is highly dependent on the cost of the house being damaged by a slide. An alternate mitigation strategy is to not carry out slope stabilization, but to protect the house from the slide by installing a barrier at a cost of (–\$20,000) so that the consequence is reduced from (–\$200,000) to (–\$50,000). The slide consequence takes into account that some damage could occur to the house, and cleaning out of slide debris and repairs to the barrier may be required. Possible types of barriers suitable for these conditions are a fence fabricated with steel netting, mechanically stabilized earth (MSE) wall, or a gabion wall (wire baskets filled with rock).

Analysis of the strategy to protect the house with a barrier is shown in Figure 7.6, the results of which are as follows.

a. The expected value of the existing condition is the same as that shown in Figure 7.5. That is, the stabilization cost is zero, the probability of a slide is 50% and the consequence of damage to the house is (–\$200,000), so the EV for this option is:

$$[EV_{existing} = ((\$0) + 0.5 \cdot (-\$200{,}000)) = (-\$100{,}000)] \\ -\ existing \quad conditions$$

b. For the barrier construction option, the probability of a slide is 50%, which is unchanged from existing conditions because no stabilization work is carried out.
c. The expected value of the option to protect the house with the barrier costing (–\$20,000) and a consequence of a slide of (–\$50,000) is:

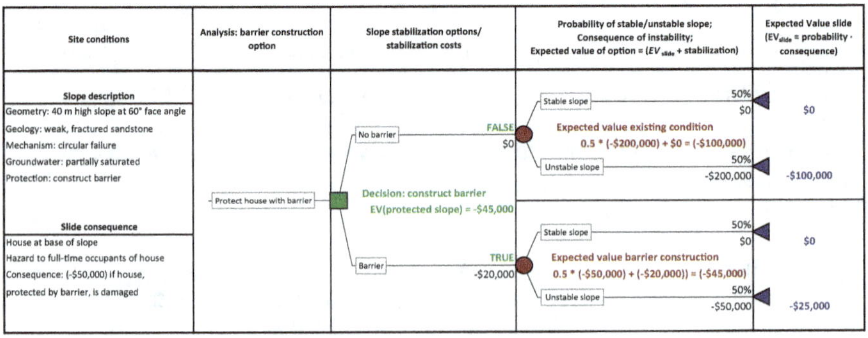

Figure 7.6 Decision analysis of option to protect the house with a barrier costing \$20,000.

$$[EV_{barrier} = (0.5 \cdot (-\$50{,}000)) + (-\$20{,}000) = (-\$45{,}000)] - \textit{barrier construction}$$

Decision analysis shows that the optimum strategy ("TRUE") is to construct the barrier because the *EV* is (–$45,000) compared to the *EV* of existing conditions of (–$100,000).

Notes of pragmatism regarding barrier construction: first, the barrier must have the capacity to withstand the impact energy of a slide, and will need to have a steep, up-slope face so that the slide will be contained and not flow/roll over the structure (Wyllie, 2014). Second, sufficient space must be available behind the house to accommodate the barrier footprint. Third, construction equipment must be able to access the site. Fourth, construction materials, such as sound rock for gabions must be readily available.

7.4 PROBABILITY IN DECISION ANALYSIS

Parameters used in decision analysis such as probability of occurrence, consequence of events and remedial measures can be expressed as discrete values that are used to calculate discrete expected values (*EV*) for events as shown in Figures 7.5 and 7.6. In reality, all these parameters have a degree of uncertainty that can be expressed most simply as ranges of values greater than and less than the most likely values, or more precisely as probability distributions.

The software PrecisionTree that was used to develop the decision trees in Figures 7.5 and 7.6 incorporates features that allow sensitivity analyses to be carried out in which the sensitivity of calculated expected values to changes in the input parameters can be determined. For example, for the option to install tie-backs shown in Figure 7.5, the cost of installing tie-back anchors has been stated as (–$50,000), with the cost of damage in the event of slope failure being (–$200,000).

The sensitivity of the Expected Value ($EV_{\text{total/tie-backs}}$) to the cost of installing the tie-backs, and the cost of damage to the house if a slide occurs can be found as follows. The estimated ranges of costs, expressed as triangular distributions – minimum cost; most likely cost; maximum cost – are as follows:

- Tie-back costs – (–$40,000); (–$50,000); (–$75,000)
- Slide damage costs - (–$175,000); (–$200,000); (–$400,000)

These ranges of costs take into account that cost estimates for geotechnical projects are often low and that cost overruns are, unfortunately, common.

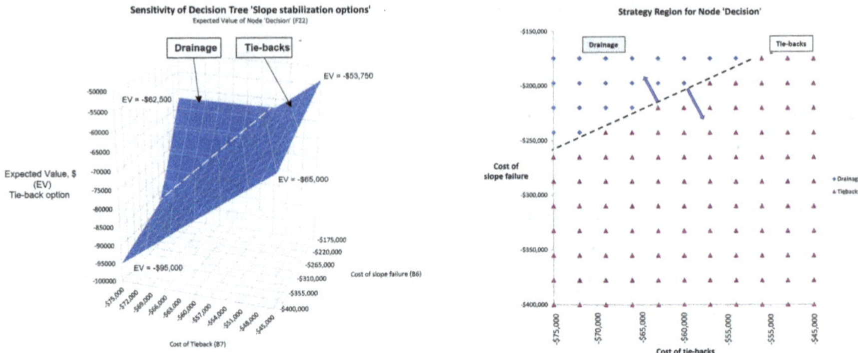

Figure 7.7 Sensitivity analysis of influence of cost of installing tie-backs, and cost of a slide, on the expected value of the decision to stabilize the slide with tie-backs, (a) Two-way sensitivity plot and (b) Strategy region plot.

If the sensitivity analysis function in PrecisionTree is run with these ranges of costs, their influence on the expected value, $EV_{\text{total/tie-backs}}$ is shown as a "two-way plot" in Figure 7.7a) where the tie-back costs and the damage costs are on the horizontal axes, and $EV_{\text{total/tie-backs}}$ is shown on the vertical axis. Note that all these costs are negative values with the most unfavourable condition being when the cost of the tie-backs is (–$75,000) and the cost of a landslide is (–$400,000) such that the $EV_{\text{total/tie-backs}}$ is (–$95,000). In comparison, when the tie-back cost is (–$45,000) and the landslide cost is (–$175,000), $EV_{\text{total/tie-backs}}$ is reduced to a more favourable value of (–$53,750). The four $EV_{\text{total/tie-backs}}$ values for the corners of the plot are shown on Figure 7.7a).

The sensitivity analysis also shows that when the cost of the tie-backs is more than (–$55,000), and the cost of a slide is under (–$250,000), then the installation of drain holes at a cost of (–$10,000) becomes the optimum strategy. The change in the mitigation strategy is indicated by the change in gradient of the plane in the Two-way plot (Figure 7.7a), and the change in the Strategy region plot for the drainage option in the upper, left corner of the plot (Figure 7.7b).

7.5 RISK MITIGATION – PRECAUTION AND RESILIENCE

The decision analyses discussed in this chapter all involve examination of the cost-benefits of possible actions, with the objective of identifying the action that minimizes the expected value of consequences such as property damage, injuries, traffic interruptions or construction schedule delays.

The total expected value of an action is the sum of the mitigation cost, and the expected value of the consequence of the event occurring.

Clearly, determination of the optimum strategy will depend on the mitigation costs for each option being maintained within defined limits. This is demonstrated in the study of slope stabilization options shown in Figures 7.5 to 7.7 where the tie-back strategy is only valid if the cost of installation is less than about (−$55,000); as the costs increase above (−$55,000), drainage becomes the optimum strategy. The value of conducting rigorous decision analysis, particularly if it incorporates sensitivity analysis, is that the influence of uncertainty in the costs of mitigation and of consequences on selection of the optimum strategy can be determined.

Additional intrinsic factors that may influence mitigation and consequence costs are **precautions** that are taken to limit consequences of damage, and the **resilience** of the system to withstand and recover from damage (Ale, Hartford, & Slater, 2021). Referring to the slope stabilization study shown in Figures 7.5 and 7.6, precautions that could be taken to minimize the risk of damage to the house and injury to the residents are to study stability conditions, carry out decision analysis to identify the optimum stabilization action, and then implement this action. In contrast, resilience will examine the probability that the slide will impact the building, capacity of the building to withstand impact, probability that it will be occupied, and the availability/capacity of emergency resources to respond to an event that damages the building.

Where precautions and resilience are considered as part of the decision analysis, monetary and ethical issues may play a part in the decision. That is, the cost of the precautionary measures may be the responsibility of the residence owner if the slide is on their land, or by government if the slide is on public land. Furthermore, if construction of the residence was approved by the government agency, then responsibility for the precautions may be borne by the government, whereas if construction was entirely the decision of the owner, then the cost of precautions should be borne by the owner.

This matter is complicated by the issue of resilience. If damage and injury occur as the result of a slide, then the government, and possibly an insurance company, may bear the entire, or portion, of the costs for rescue, medical expenses and recovery from the accident, in addition to the overhead of providing these services. If the hazard has been identified and considered to be of concern, is it ethical to allow the residents to be unprotected from the slide? Furthermore, who is responsible for enforcing and paying for remedial work on the residence owner's land if the owner is unwilling to take action?

As shown in Figure 7.3, both direct costs (damage) and indirect costs (providing rescue and medical services, and possible litigation) need to be considered in identifying remedial costs.

7.6 CASE STUDIES

The application of decision analysis to the four case studies discussed in the previous chapters is presented below. These four cases include numeric values for costs of damage and remedial work, and probabilities of event occurrence, to illustrate applications of decision analysis. These models can be readily adapted to suit similar project conditions.

7.6.1 Debris flow dam

If a debris flow hazard exists above a residential development, a decision may be required on the optimum method of protecting the houses and residents. The three basic decisions are to compare current conditions with two options of either constructing a debris flow dam with the capacity to contain potentially hazardous events, or evacuating houses in the debris flow runout area. Factors that need to be considered in this decision are the possible frequency and magnitude of future debris flows, and whether a dam can be constructed with the capacity to contain these events, taking into account the uncertainty in predicting their magnitude. Alternatively, protection could be provided by either prohibiting development in the hazard area if this is a proposed development, or relocating existing houses. These decisions will also take into account uncertainty in the size of the hazard area.

Figure 7.8 shows the decision tree defining the three options, assuming that houses are already located on the debris flow runout fan. The decisions comprise the following steps:

- **Remediation options and cost** - estimated remediation costs are: (–$200,000) for constructing the dam, (–$2 million) for relocating the houses and ($0) for maintaining existing conditions.
- **Debris flow probability** - a debris flow over the life of the project is estimated, based on past records, to have a 60% chance of occurrence. This probability is independent of the actions to build the dam or move the houses because no remedial action has been taken in the source area.
- **Consequences**- possible damage to the houses if the debris flow occurs can be estimated. If the dam has been constructed, the probability that the dam will be overtopped is 2%, with a probability of 98% that all material will be contained. If the dam is overtopped, the probability that damage will occur is 25%, and the estimated cost to remove the accumulated debris and carry out minor repairs is (–$50,000). For existing conditions, the possibility of damage is 80%, and damage caused by the event is estimated to cost (–$750,000). However, if the houses are relocated, it is estimated that the probability of damage by debris flows is 70%, but the cost of this damage to remaining infrastructure is (–$20,000).

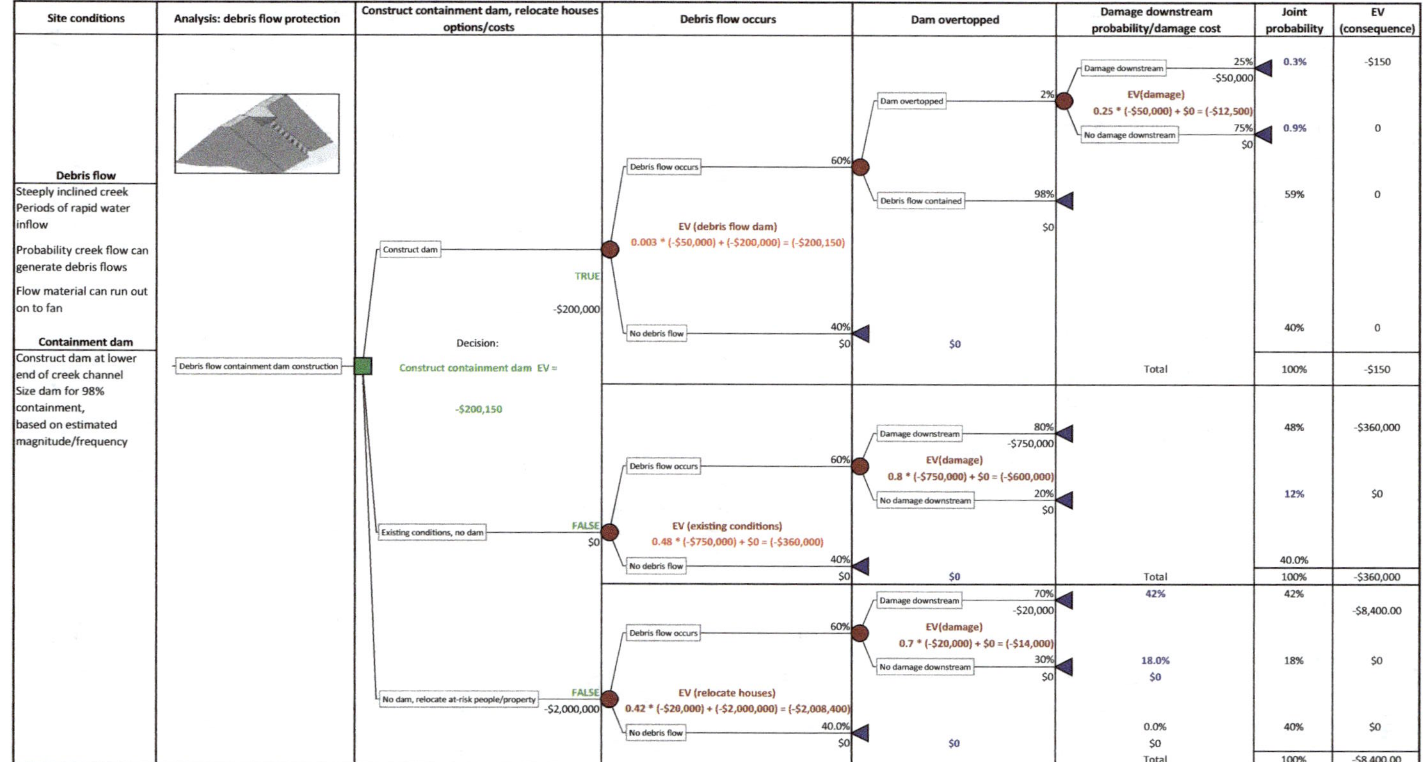

Figure 7.8 Decision tree illustrating analysis of options to protect houses from debris flows by either constructing a containment dam, or by relocating the houses; protection options are compared to existing conditions (plot by PrecisionTree, Lumivero Corp.)

Calculation of the Expected Values (EV) for each decision shows the following (Figure 7.8):

- **Construct dam cost** = (–\$200,000) – if a debris flow occurs (P=60%), the dam is overtopped (P=2%) and damage of (–\$50,000) occurs downstream at (P=25%), the joint probability of this event is:

  Joint probability of damage = [(0.6)·(0.02)·(0.25) = 0.3%].

  Expected value for the option of constructing the containment dam is calculated for each branch from the joint probabilities and consequences of debris flow events as follows:

  $$EV_{damage} = (0.003 \cdot (-\$50{,}000)) \qquad = (-\$150)$$

  $$EV_{no\ damage} = (0.009 \cdot (\$0)) \qquad = (\$0)$$

  $$\mathbf{EV_{damage\ dam}} \qquad = (\mathbf{-\$150})$$

  The total expected value is equal to the cost of constructing the dam (–\$200,000) plus EV_{damage}:

  $$[EV_{protection\ dam} = (-\$200{,}000) + (-\$150) = (\mathbf{-\$200{,}150})].$$

- **Existing conditions** - if a debris flow occurs (P=60%), and unprotected houses on the run-out fan are damaged at a cost of (–\$750,000) at (P=80%), joint probability is:

  Joint probability of damage = [(0.6) · (0.8) = 48%]

  Expected value of damage for existing conditions is:

  $$EV_{damage} = (0.48 \cdot (-\$750{,}000)) \qquad = (-\$360{,}000)$$

  $$EV_{no\ damage} = (0.12 \cdot (\$0)) \qquad = (\$0)$$

  $$\mathbf{EV_{damage\ existing}} \qquad = (\mathbf{-\$360{,}000})$$

  Because the remediation cost is zero, the total expected value is:

  $$\left[EV_{existing\ conditions} = (\mathbf{-\$360{,}000})\right].$$

- **Relocate houses** - if a debris flow occurs (P=60%), and remaining infrastructure on the run-out fan is damaged at a cost of (–\$20,000) at (P=70%), joint probability is:

$$Joint\ probability\ of\ damage = [(0.6)\cdot(0.7) = 42\%]$$

Expected value of damage is:

$$EV_{damage} = (0.42 \cdot (-\$20{,}000)) \qquad = (-\$8{,}400)$$

$$EV_{no\ damage} = (0.18 \cdot (\$0)) \qquad = (\$0)$$

$$\boldsymbol{EV_{damage\ house\ relocation}} \qquad = (-\$8{,}400)$$

The total expected value is equal to the cost of relocating the houses (−$2 million) plus the EV_{damage}:

$$\left[EV_{relocate\ houses} = \left(-\$2\ million\right) + \left(-\$8{,}400\right) = \left(-\$2{,}008{,}400\right)\right].$$

This analysis shows that the optimum decision (TRUE) is to construct the dam at an expected value of (−$200,150). That is, the cost of constructing the dam is offset by the low probability and low expected value of damage to the houses in the event of a debris flow that is almost entirely contained by the dam. In contrast, the total expected value of damage for existing conditions with no dam (−$360,000), as well as the expected value of relocating the houses (−$2,008,400) are substantially more than that for the dam construction option. This analysis accounts for uncertainty in both the frequency and magnitude of future debris flows by allocating probability values to the occurrence of events.

An illustration of a decision related to debris flow hazards is a plan to protect an existing residential area, and other critical infrastructure including a highway and railway, located on the Cougar Creek debris flow fan in the town of Canmore, Alberta in Canada. A debris flow in 2013 caused about $40 million damage, and in this situation, it was decided to construct a 33 m high containment dam above the residential area (Jakob, Weatherly, Bale, & Perkins, 2017).

7.6.2 Rock falls

Highways and railways located in mountainous terrain may be subject to rock fall hazards originating from both natural slopes and excavated faces. Such events can result in traffic interruptions, damage to infrastructure and vehicles, and injury to persons, all of which will be a cost to the transportation system. Mitigation of these hazards can be achieved by stabilization of the slope to prevent falls from occurring and/or constructing protection measures such as fences, barriers and ditches.

Decision analysis can be used to study the cost-benefits of mitigating the rock fall hazards and reducing the probability of accidents and their costs. Figure 7.9 shows a decision tree comparing the expected value of rock falls

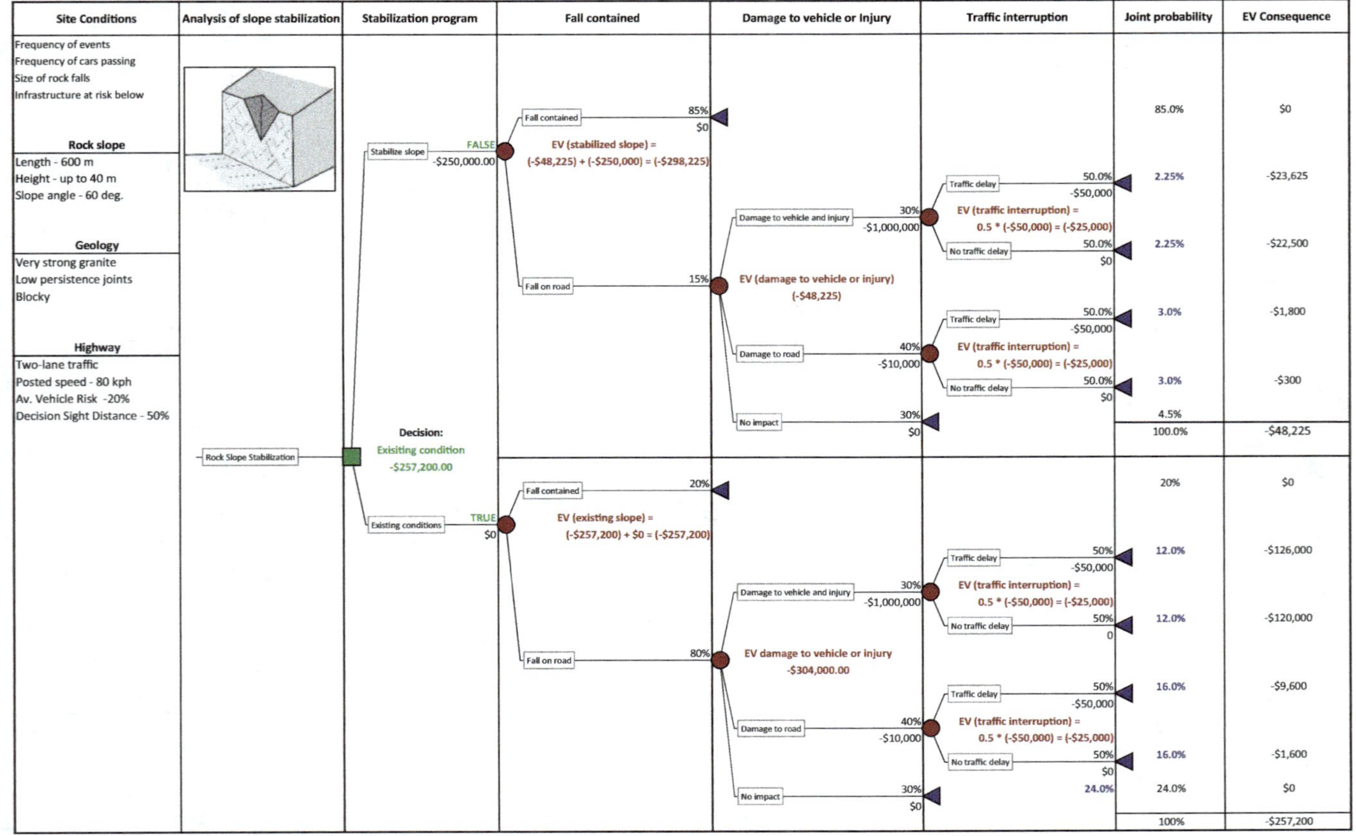

Figure 7.9 Decision tree illustrating analysis of rock slope stabilization program on a highway.

for existing conditions to those for a stabilized slope. Steps in the decision analysis are as follows:

- **Mitigation options and costs** - the two options at the decision point are to implement a slope stabilization program at a cost of (–\$250,000), or to accept existing conditions at a cost of (\$0).
- **Consequences** – to examine the probability and consequence of slope instability, records that document previous falls can be studied. These records show that 80% of previous falls reached the road surface, and that the other 20% were contained in the catchment area at the base of the slope. For falls that reach the road surface, records show that 30% did not result in damage or traffic interruptions, mainly because of their small size due to shattering on impact. Of the remaining 70% of the falls, 40% only damaged the road requiring about (–\$10,000) of repairs, while 30% of falls resulted in damage to vehicles and injury to persons, at an average cost of (–\$1 million) per event. For each fall that lands on the road, the possibility of interruptions to traffic will occur at a probability of 50%, at an average cost of (–\$50,000) per event.

Other issues to consider regarding both future probabilities and consequences of falls are that falls may become more frequent in the future because of climate change and higher rainfall, and the consequences may be more severe because of increases in traffic.

Calculation of the *EV*s for the two decisions shows the following values (Figure 7.9):

- **Existing conditions** – the lower half of the decision tree in Figure 7.9 shows probabilities and consequences for existing conditions. On the tree, the consequences of rock falls for existing conditions are identical to those for the stabilized slope because the stabilization work has no effect on the operating conditions of the highway. The effect of the stabilization work is to reduce the probability of falls that reach the road from ($P=80\%$) for existing conditions to ($P=15\%$) for the stabilized slope. For existing conditions, a typical event is one that falls on the road ($P=80\%$), causes damage to vehicles and injuries ($P=30\%$) costing (–\$1 million) and results in traffic interruptions ($P=50\%$) costing (–\$50,000). The joint probabilities for existing conditions are:

$$\textit{Joint probability}\left(\textit{damage to vehicle, traffic interruption}\right) = (0.8 \cdot 0.3 \cdot 0.5) = 12\%.$$

$$\textit{Joint probability}\left(\textit{damage to road, traffic interruption}\right) = (0.8 \cdot 0.4 \cdot 0.5) = 16\%.$$

The expected value of traffic interruptions, injury and damage is calculated for each branch from the joint probabilities and consequences of falls as follows:

$$[EV_{traffic/damage/injury} = (0.12 \cdot ((-\$(50{,}000) + (-\$1{,}000{,}000))) = (-\$126{,}000)]$$

$$[EV_{damage/injury} = (0.12 \cdot ((\$0) + (-\$1{,}000{,}000))) = (-\$120{,}000)]$$

$$[EV_{traffic/road} = (0.16 \cdot ((-\$(50{,}000) + (-\$10{,}000))) = (-\$9{,}600)]$$

$$[EV_{road} = (0.16 \cdot ((\$(0) + (-\$10{,}000))) = (-\$1{,}600)]$$

$$EV_{existing\ conditions} = (-\$257{,}200)$$

The expected value of existing conditions, with no stabilization work carried out, is:

$$EV_{existing\ conditions} = (-\$257{,}200) + (\$0) = (-\$257{,}200)$$

- **Stabilized slope** – after a slope stabilization program costing (–\$250,000), the probability of a fall on the road is reduced to 15%, compared with the probability of a fall on the road of 80% for existing conditions. After stabilization, a typical event is one that falls on the road (P=15%), causes damage to vehicles and injuries (P=30%) costing (–\$1 million) and results in traffic interruptions (P=50%) costing (–\$50,000). The joint probabilities for the stabilized slope are:

$$Joint\ probability\ (damage\ to\ vehicle,\ traffic\ interruption) = (0.15 \cdot 0.3 \cdot 0.5) = 2.25\%.$$

$$Joint\ probability\ (damage\ to\ road,\ traffic\ interruption) = (0.15 \cdot 0.4 \cdot 0.5) = 3\%.$$

The expected value of traffic interruptions, injury and damage is calculated for each branch from the joint probabilities and consequences of falls as follows:

$$[EV_{traffic/damage/injury} = (0.0225 \cdot ((-\$(50{,}000) + (-\$1{,}000{,}000))) = (-\$23{,}625)]$$

$$[EV_{damage/injury} = (0.0225 \cdot ((\$0) + (-\$1{,}000{,}000))) = (-\$22{,}500)]$$

$$[EV_{traffic/road} = (0.03 \cdot ((-\$(50{,}000) + (-\$10{,}000))) \quad = (-\$1{,}800)]$$

$$[EV_{road} = (0.03 \cdot ((\$(0) + (-\$10{,}000))) \quad = (-\$300)]$$

$$EV_{damage} \quad = (-\$48{,}225)$$

The expected value for the stabilized slope is the sum of the EV for the consequences of the rock falls and the stabilization cost as follows:

$$EV_{st\,abilized\ slope} = (-\$48{,}225) + (-\$250{,}000) = (-\$298{,}225)$$

The analysis shows that a stabilization program costing (–$250,000) is possibly not justified because the expected value of falls for the existing slope (–$257,200) is slightly less (TRUE) than the expected value of falls for the stabilized slope of (–$298,225). Further analysis of the decision tree would be instructive on finding an optimum stabilization program.

7.6.3 Tunnel stability

When excavating a tunnel, unfavourable conditions may be encountered such as seams of weak, fractured rock and/or zones of high water inflow. Such conditions may require installation of extra support of the loose rock and grouting to control water inflow; this extra work could result in cost overruns and schedule delays. These consequences can be particularly severe, and dangerous to the miners, if the conditions are unexpected.

Information on advanced geological conditions can be obtained by drilling probe holes into the face to identify possible seams of weak rock and high groundwater inflow ahead of the excavation. While probe holes can provide information that can be used to plan ground support and water control, the holes will be expensive and cause schedule delays. The cost-benefit of drilling the holes can be assessed using decision analysis where the cost of the holes can be compared with the expected value of extra remedial work, taking into account the degree of uncertainty in the geology.

Figure 7.10 shows a decision tree that analyzes the cost-benefit of drilling probe holes comprising the following steps:

Probe hole cost – it is estimated that the cost of drilling a series of probe holes as the tunnel progresses is (–$20,000).

Expected adverse geology – mapping of rock outcrops in the tunnel area shows that the tunnel may intersect single faults, or even multiple faults, and that these faults may be sources of water inflow. The probability of occurrence of these features is estimated at 20% for multiple faults, 55% for a single fault and 25% for no faults.

Consequence of intersecting faults – if the tunnel intersects a fault, costs will be incurred to stabilize the excavation and control water inflow, and these costs will be greater if the faults are intersected unexpectedly than if the probe holes provide a warning of these hazards. If the

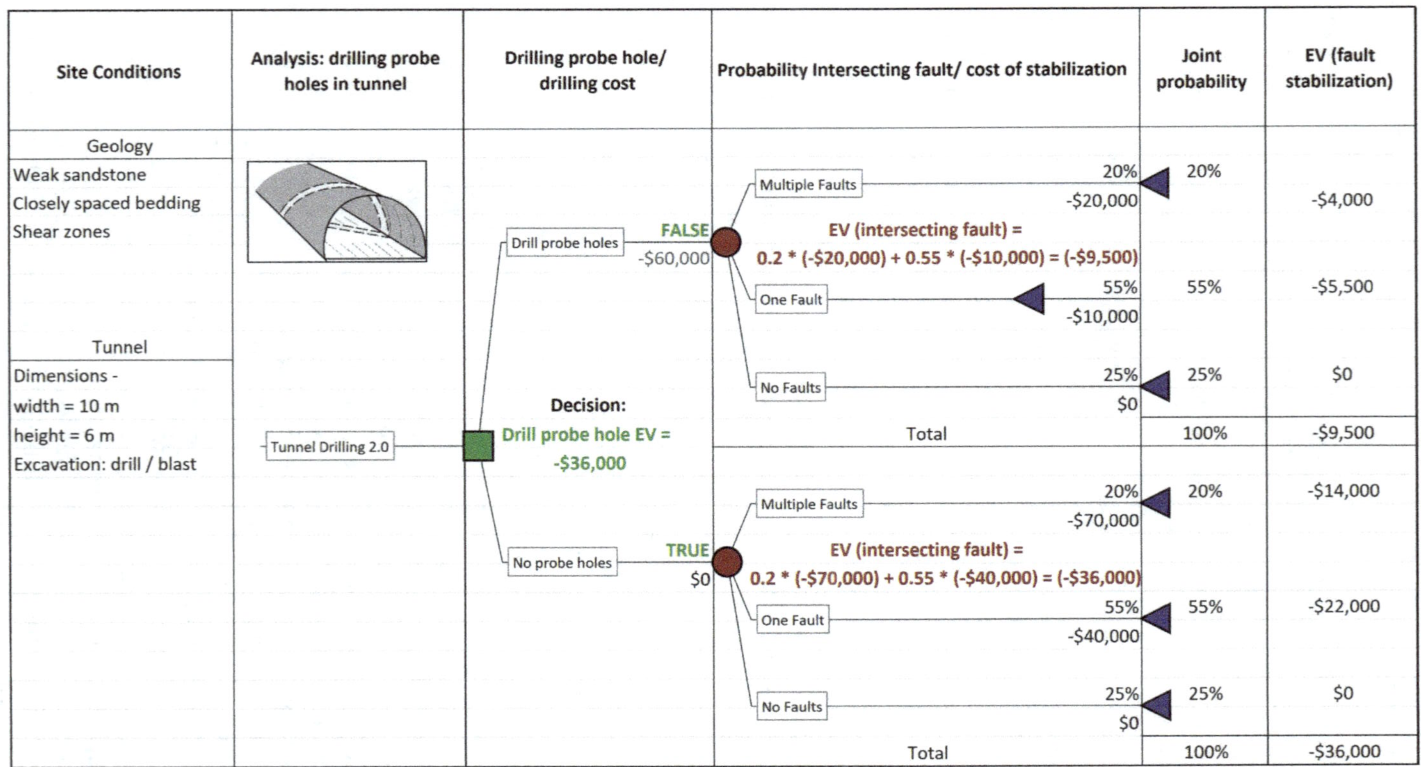

Figure 7.10 Decision tree illustrating analysis of drilling probe holes ahead of tunnel face to obtain advanced geological information.

probe holes identify a fault ahead of the face, precautions could include reducing the advance length and installing support close to the face, and grouting the rock around the tunnel. Costs for the stabilization work are shown on the decision tree in Figure 7.10.

Calculations of the EV for the two options are as follows (Figure 7.10):

- **No probe holes** – the probability of intersecting multiple faults is estimated at 20%, and intersecting only one fault is 55%. These probabilities are the same whether or not the probe holes are drilled because the geology is independent of the drilling operation. The expected values for stabilizing the faults are:

$$EV_{multiple\ faults} = (0.20 \cdot (-\$70,000)) = (-\$14,000)$$

$$EV_{single\ fault} = (0.55 \cdot (-\$40,000)) = (-\$22,000)$$

$$EV_{fault\ stabilization} = (-\$36,000)$$

The expected value for the option of not drilling probe holes is equal to EV for the consequences of intersection faults with no cost for drilling probe holes as follows:

$$EV_{no\ probe\ holes} = (-\$36,000) + (\$0) = (-\$36,000)$$

- **Probe holes** – drilling probe holes will reduce support costs, compared to the option of not drilling probe holes, because support can be installed ahead of the face if the probe holes identify faults. The expected value of stabilizing faults ahead of the face is calculated as follows:

$$EV_{multiple\ faults} = (0.20 \cdot (-\$20,000)) = (-\$4,000)$$

$$EV_{single\ fault} = (0.55 \cdot (-\$10,000)) = (-\$5,500)$$

$$EV_{fault\ stabilization} = (-\$9,500)$$

The Expected Value for the option of drilling probe holes is the sum of the EV for the consequences of intersection faults and drilling probe holes as follows:

$$EV_{probe\ holes} = (-\$9,500) + (-\$20,000) = (-\$29,500)$$

The analysis shows that the cost of drilling the probe holes is justified because of the reduction in the cost of supporting the faulted rock from (–\$36,000) to (–\$29,500) if the locations of the faults ahead of the face are identified by the probe holes.

7.6.4 Dam foundation

The capacity of concrete dams to withstand both sliding and overturning forces of the contained water depends on the mass of the structure, and the shear resistance of the foundation between the concrete and the rock. The dam must also be stable against forces generated by seismic ground motions. In some cases, it has been necessary to improve the stability of existing dams to account for seismic design parameters that were not considered in the dam's initial design. For concrete gravity dams, additional support has been developed by installing tensioned rock anchors through the dam and anchored into the foundation. The tensioned rock anchors act to provide resistance to to sliding by increasing the normal force on the concrete/rock interface, as well as counteracting the buoyancy (uplift) forces from seepage beneath the dam.

The required anchors are usually multi-strand cables installed in holes drilled at a precise alignment with a diameter that may be as large as 300 mm (12 in.) and installed deep in the foundation rock to create a stable, cement-grouted bond zone. After installation, the anchors are tensioned with a hydraulic jack and then locked off at a long-term, design working load. These demanding requirements for anchor installation means that high-quality construction is essential, and a decision may be made to install test anchors to optimize the drilling and anchoring methods.

Figure 7.11 shows a decision tree of a cost-benefit analysis for a test anchor program to investigate the drilling and grouting methods, involving the following steps.

- **Decision options** – test program costs (–$15,000); the value of testing is the expectation that the tests will reduce the risks of the production holes being misaligned, and of grout leakage in the bond zone that will result in poor bonding of the anchors.
- **Probability of drilling and grouting success** – previous experience with similar projects shows that the estimated probabilities of drill hole misalignment is 60%, and of grout leakage in the bond zone is 20%. The test anchors should reduce the probability of hole misalignment to about 20%.
- **Consequences** – again previous experience shows that the cost of a misaligned drill hole could be (–$100,000) because of the need to seal and redrill the hole, while the cost of grout leakage could be (–$10,000) to seal the hole and prevent leakage. It is expected that the probability of grout leakage will increase if the hole is misaligned and is outside the zone of consolidation grouting in the foundation.

The expected values for the two options are as follows (Figure 7.11).

- **Install test anchors** – if test anchors are installed, the costs for drill hole misalignment and grout leakage are the same as that for not installing test anchors, but the test drilling reduces the probability of these events occurring.

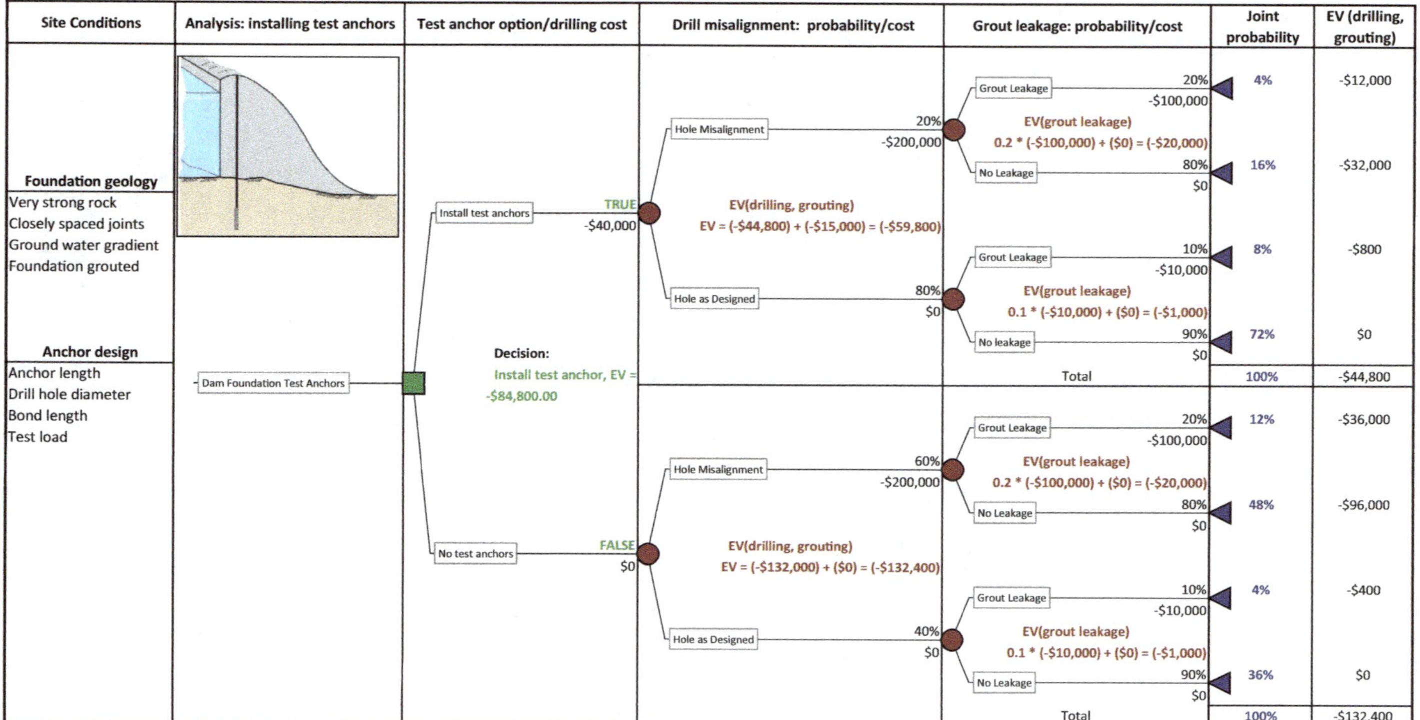

Figure 7.11 Decision tree illustrating analysis of drilling test holes for anchors required to stabilize a dam against sliding when subject to seismic ground motions.

The Expected Value of redrilling anchor holes and grouting the bond zone is calculated for each branch from the joint probabilities and consequences of hole misalignment as follows:

$$
\begin{aligned}
[EV_{grout/redrilling} &= (0.04 \cdot ((-\$(100{,}000) + (-\$200{,}000))) &&= (-\$12{,}000)]\\
[EV_{redrilling} &= (0.16 \cdot ((\$0) + (-\$200{,}000))) &&= (-\$32{,}000)]\\
[EV_{grouting} &= (0.08 \cdot ((-\$(10{,}000) + (\$0))) &&= (-\$800)]\\
[EV_{no\ grout/hole\ aligned} &= (0.72 \cdot (\$(0)) &&= (\$0)]\\
\hline
EV_{grouting/drilling} & &&= (-\$44{,}800)
\end{aligned}
$$

The expected value of the test anchor installation program is:

$$EV_{\text{existing conditions}} = (-\$44{,}800) + (-\$15{,}000) = (-\$59{,}800)$$

- **No test anchors** – if test anchors are not installed, then for the production drill holes, the probability of hole misalignment may be greater than if the test anchors had been installed and the drilling procedures verified. If hole misalignment does occur, then the probability of having to grout the bond zone will increase.

The expected value of redrilling anchor holes and grouting the bond zone is calculated for each branch from the joint probabilities and consequences of hole misalignment as follows:

$$
\begin{aligned}
[EV_{grout/redrilling} &= (0.12 \cdot ((-\$(100{,}000) + (-\$200{,}000))) &&= (-\$36{,}000)]\\
[EV_{redrilling} &= (0.48 \cdot ((\$0) + (-\$200{,}000))) &&= (-\$96{,}000)]\\
[EV_{grouting} &= (0.04 \cdot ((-\$(10{,}000) + (\$0))) &&= (-\$400)]\\
[EV_{no\ grout/hole\ aligned} &= (0.36 \cdot (\$(0)) &&= (\$0)]\\
\hline
EV_{grouting/drilling} & &&= (-\$132{,}400)
\end{aligned}
$$

The expected value of existing conditions, with no test anchors is:

$$EV_{existing\ conditions} = (-\$132{,}400) + (\$0) = (-\$132{,}400)$$

The decision analysis in Figure 7.11 indicates that installation of test anchors is of value ("TRUE") because verification of the drilling procedures will reduce the probability of hole misalignment from $P=60\%$ to $P=20\%$, and the expected value of anchor installation is reduced from (–\$132,400) to (–\$59,800).

Chapter 8

Reliability Design of geotechnical structures

Factors of safety commonly used in geotechnical engineering are based on experience, and an extensive body of project work that has provided practitioners with guidelines on factor of safety values that are considered to be generally acceptable (see Section 2.1.2 and Table 2.1). However, it is common to use the same factor of safety for each type of application such as factor of safety=1.3–1.5 for slopes and factor of safety=3–5 for seepage, regardless of the degree of uncertainty in the design. The use of "standard" factors of safety may be because of tradition or regulation (Phoon and Ching, 2015; Chapter 3, Duncan and Sleep)

With the objective of using factors of safety that are more consistent with uncertainty in design parameters, reliability calculations provide a means of evaluating the combined effects of uncertainties, and of distinguishing between conditions where uncertainties are particularly high or low (Wu, Tang, Sangrey, & Baecher, 1989). "Reliability" as it is used in reliability theory is the probability of an event occurring, or the probability of a "positive outcome". Reliability is the complement of probability of failure. That is, the greater the reliability index, the lower the probability of failure (Phoon, 2016).

This chapter discusses both simple methods of evaluating reliability without using more data, time or effort than is used in conventional geotechnical design (see Section 8.2), and more sophisticated reliability analysis in which design parameters have different levels of uncertainty are defined by probability distributions such as Beta and Lognormal. Also discussed is Reliability Based Design (RBD) where the structure is designed to a specified reliability value (see Section 8.3).

8.1 SELECTED STATISTICAL TERMS

The following is a brief discussion on selected statistical terms that have application to reliability analysis.

DOI: 10.1201/9781003271864-8

8.1.1 Correlated and uncorrelated variables

Reliability analysis can consider that variables such as the shear strength parameters cohesion (c) and friction angle (ϕ) are either correlated or uncorrelated. Correlation is defined by the correlation coefficient (r) where r is positive if, for example, c increases with increasing values of ϕ, and is negative if c decreases with increasing values of ϕ. Also, the two parameters are perfectly correlated ($r=1$) if each value of c relates to a specific value of ϕ, and $r=0$ for uncorrelated parameters.

Correlated design parameters are not commonly used in design because it is rare to find correlation in geotechnical parameters. For example, if the friction angle of a soil was measured, it is unlikely that this would correspond to a specific correlated value of cohesion unless specific data on the soil properties were available. The Reliability Based Design (RBD) method described in Section 8.3 incorporates correlated c and ϕ values.

8.1.2 Coefficient of variation (COV)

While standard deviation is a useful indicator of the amount of scatter in a variable, the degree of dispersion is easier to see in context if it is expressed in terms of the coefficient of variation (COV), which is defined by equation (8.1):

$$COV = \frac{\sigma}{\mu} \tag{8.1}$$

where σ is the standard deviation and μ is the mean value of the data set. Equation (8.1) shows that the COV increases as the standard deviation, and the dispersion in the data increases.

The COV is a dimensionless measure of the amount of scatter and is usually expressed as a percentage, with the COV value increasing as the degree of scatter, and the standard deviation, increases.

For reference, Table 2.2 (Chapter 2) lists the COV values for 17 soil parameters determined from laboratory and *in situ* tests. The quoted COV values vary from 3% to 7% for unit weight, which is a parameter that can be readily measured, to COV=15% to 45% for standard penetration tests using blow counts that usually have a margin of error, to COV=130% to 240% for coefficient of permeability for partially saturated clay that is indicative of the difficulty in measuring the very low permeability of clay.

8.1.3 Probability of failure

If the factor of safety of a geotechnical structure has been calculated using probability distributions for parameters such as the shear strength and location of the sliding surface, the calculated factor of safety will also be defined by a probability distribution.

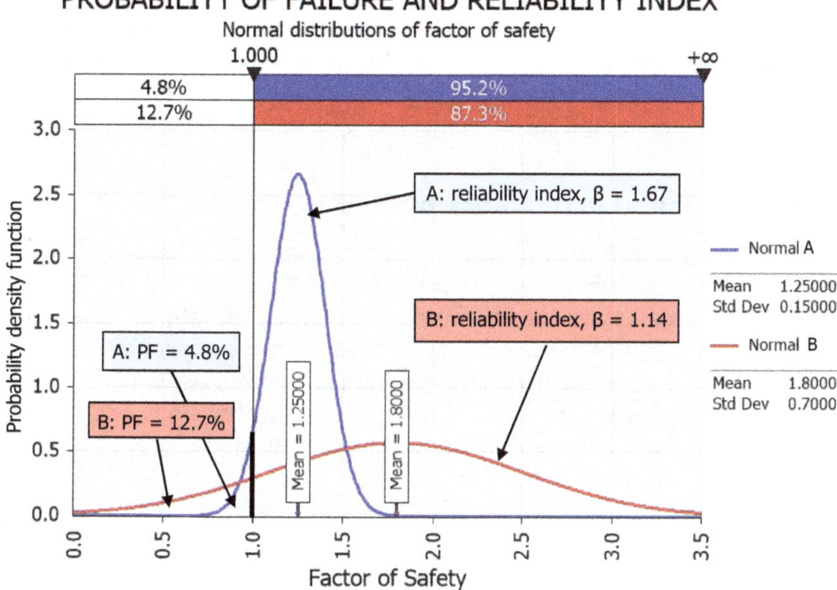

Figure 8.1 Normal distributions A and B of factor of safety showing average, standard deviation, probability of failure and reliability index for each distribution; reliability index is discussed in Section 8.2 below (plot generated by @Risk, Lumivero Inc.).

In Figure 8.1, the normal distributions of the factor of safety have the following mean (μ) and standard deviation (σ) values:

Curve A: $\mu_A=1.25$, $\sigma_A=0.15$; Curve B: $\mu_B=1.8$, $\sigma_B=0.7$

The difference in the level of uncertainty, and standard deviation between the two factor of safety distributions is indicated by the width of the plots, with plot B having more uncertainty, greater standard deviation and a wider plot than plot A. For both plots, the left side limit extends beyond the factor of safety=1.0 to values of factor of safety<1, indicating that failure is possible. The area under each curve where factor of safety<1.0, compared to the total area under the curve that has a value of 1.0, defines the probability of failure (*PF*) of the slope. Therefore, for curve A the $PF=4.8\%$ and for curve B the $PF=12.7\%$. This result shows that although Plot B has the higher average factor of safety value of $FS=1.8$ compared to Plot A with an average factor of safety of $FS=1.25$, Plot B has a higher probability of failure because of the greater uncertainty in the calculated values.

The example shown in Figure 8.1 has used normal distributions because they are commonly used for statistical data analysis. However, normal distributions have disadvantages for geotechnical applications because they extend to infinity in both directions giving negative values at the lower end, and because the plots are symmetrical. These disadvantages are overcome by using lognormal distributions that extend from zero to infinity with no negative values, or Beta distributions that have defined upper and lower limits. Both these distributions can be skewed to the left or right to account for low or high data concentrations.

8.1.4 Meaning of probability of failure

Figure 8.1 shows how the area under the factor of safety distribution curve that is less than 1.0 represents the probability of failure of the slope. The following is a discussion on the meaning of the term "probability of failure".

If the components of a geotechnical system comprise capacity (resistance) - C such as the shear strength on the sliding plane, and the demand - D such as the weight of the sliding mass, water pressure in the slope and external loads, then the factor of safety is defined by the ratio:

$$FS = \frac{C}{D} \tag{8.2}$$

where the factor of safety will be less than 1 if $(D > C)$.

Alternatively, the stability of the structure can be defined by the difference between C and D, with failure occurring if $((C - D) < 1.0)$. If the stability analysis calculates normal distributions for the Capacity and Demand forces for the slope, then failure may occur if portions of the two distributions for C and D overlap, as shown in Figures 2.3 and 8.3.

Calculation of the "probability of failure" using the ratio (C/D), or the difference $(C - D)$ is a mathematical function that has no time component, just as the factor of safety has no time component. That is, the calculated "probability of failure" is a number analogous to the factor of safety. Determination of temporal (annual) probability of failure requires time information on the occurrence of previous events such as the number of slope failures over a given number of years. This calculated annual probability of slope instability can then be compared, for example, with similar statistics for other areas to compare slope hazards, and plan mitigation measures.

Section 2.4 gives examples of actual annual probabilities of failure for talus slopes and slopes on transportation system calculated from records of instability.

Another issue with the term probability of failure is the meaning of "failure". Geotechnical design has the objective of achieving satisfactory performance such as tolerable movement or settlement. Where excessive settlement of a bridge occurs, for example, such that traffic can no longer safely use the bridge at design speeds, the situation would be considered one of unsatisfactory performance, although the bridge foundations would not have failed. The issue of performance and failure is addressed in limit states design as discussed in Section 2.6.

8.2 RELIABILITY INDEX

An additional method of incorporating uncertainty into geotechnical design is to use the Reliability Index, β that is the number of standard deviations between the most likely value of the factor of safety and a factor of safety of 1.0. The use in design of either the Reliability Index or the probability of failure is possible because the two terms are mathematically connected as shown in equation (8.3). The term probability of failure may be preferred because it is somewhat more intuitive in relation to the performance of a structure compared to the Reliability Index.

8.2.1 Relationship between reliability index and probability of failure

The unique relationship between the Reliability Index, β and the probability of failure, *PF* is given by equation (8.3):

$$PF = \left(1 - \phi(\beta)\right) \tag{8.3}$$

where $\phi(\beta)$ is the normal distribution of the Reliability Index. The relationship between β and the *PF* can be found using the Excel function NORMDIST(x, μ, φ, TRUE) where TRUE indicates that the cumulative form of the distribution will be used. With reference to Figure 8.1, if the Reliability Index, β=1.67 for distribution A, the equivalent value of the probability of failure is:

$$PF = \ (1 - NORMDIST\left(1.67, 0, 1, TRUE\right) \ = \ 4.8\%.$$

and for distribution B where β=1.14, *PF*=12.7%. That is, for lower reliability index values, the probability of failure increases.

Table 8.1 lists values for the relationship between β and *PF* where β is defined by a normal distribution.

While equation (8.3) and Table 8.1 show the relationship between the probability of failure, *PF* and the Reliability Index β when values of β are defined by a normal distribution, it is also possible to relate β and *PF* when β is defined by a triangular or lognormal distribution. For values of $\beta \leq 2$ approximately, little difference exists between the *PF*/β relationships regardless of which probability distribution is used; for $\beta > 2$, the divergence between *PF*/β relationships is significant with equation (8.3) giving a higher probability of failure than the triangular and lognormal distributions (Baecher & Christian, 2003).

Table 8.1 Relationship between reliability index β and probability of failure *PF* based on equation (8.3)

Reliability index, β_{normal}	*Probability of failure, PF (%)*
0.5	31
1.0	16
1.5	6.7
2.0	2.3
2.5	0.62
3.0	0.13
4.0	0.003
5.0	0.00003

8.2.2 Calculation of reliability index, β

Figure 8.1 shows plots of two normal distributions of the factor of safety defined by their means and standard deviations, and the reliability index for each plot: $\beta_A=1.67$ and $\beta_B=1.14$. These plots assume that the calculations of the factors of safety used probability distributions for the parameters defining the factor of safety (see Section 8.2.3 below on calculating the mean and standard deviation of Capacity and Demand distributions).

For a normal distribution defined by the most likely value of the mean, μ_{MLV} and standard deviation σ, the Reliability Index, β is given by equation (8.4):

$$\beta = \frac{(\mu_{MLV} - 1)}{\sigma} \tag{8.4}$$

For distribution A in Figure 8.1, where $\mu_{MLV}=1.25$ and $\sigma_A=0.15$, the Reliability Index is:

$$\beta_A = (1.25 - 1) / 0.15 = 1.67$$

Similarly, for distribution B where $\mu_{MLV}=1.8$ and $\sigma_B=0.7$, the Reliability Index is:

$$\beta_B = (1.80 - 1) / 0.70 = 1.14.$$

The reliability index can be demonstrated graphically using cumulative plots of the normal distributions for the data in distributions A and B, as shown in Figure 8.2. In these plots, the distance between a factor of safety of 1.0 and the factor of safety for each cumulative curve is equal to the product of the reliability index and the standard deviation. For example, for distribution A where $\beta_A=1.67$ and $\sigma_A=0.15$, the distance between the $FS=1.0$ and $FS_A=1.25$ is:

$$Distance = (\beta_A \cdot \sigma_A) = (1.67 \cdot 0.15) = 0.25$$

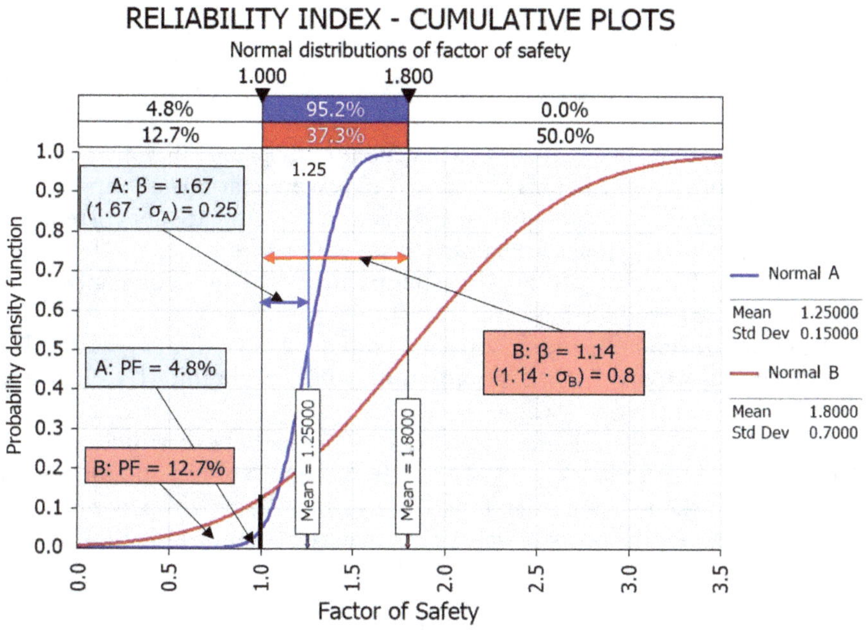

Figure 8.2 Cumulative plots of normal distributions A and B showing dimensions of reliability indices (β), and probability of failure (*PF*).

Similarly, for distribution B,

$$Distance = (\beta_B \cdot \sigma_B) = (1.14 \cdot 0.70) = 0.8$$

These two distances are plotted in Figure 8.2 where the ratio of the two lengths is: (0.8/0.25=3.2). This ratio can be measured on the plot.

Figure 8.2 shows that distribution A has a higher reliability (i.e., less dispersion and less uncertainty) than distribution B because of the shorter distance of the FS_A=1.25 mean line from the FS=1.0 line.

8.2.3 Reliability index for capacity and demand forces

For conditions where the project data is expressed as probability distributions of the Capacity, C and Demand, D, each of which are defined by means μ_C, μ_D and standard deviations σ_C, σ_D, it is possible to calculate new distributions for the mean of the margin of safety, μ_{mean}, and for the Reliability Index, β using equations (8.5) and (8.6):

$$\mu_{mean} = (\mu_C - \mu_D) \tag{8.5}$$

and

$$\beta = \frac{(\mu_C - \mu_D)}{\sqrt{(\sigma_C^2 + \sigma_D^2)}} \tag{8.6}$$

Note that equations (8.5) and (8.6) are valid for any probability distribution such as lognormal and Beta. However, these two equations are applicable only for the usual condition where Capacity and Demand are not correlated.

Figure 8.3 shows typical probability distributions for Capacity (beta distribution) and Demand (normal distribution) forces. The following parameters define the two distributions:

Demand - normal: mean, μ_D=500 kN, standard deviation, σ_D=200 kN.

Capacity - Beta: a_1=5, a_2=2.4, minimum=600, maximum=1,100, mean, μ_C=938 kN, standard deviation, σ_C=81 kN.

Parameters a_1 and a_2 define the symmetry of the Beta distribution plots that allow the plot to be skewed, within the fixed maximum and minimum values, to suit the values of the data. In this case, the plot is skewed to the right showing that the probability of the capacity being towards the low end of the 600 to 1,100 kN range is minimal.

Applying these values to equations (8.5) and (8.6), the mean margin of safety, μ_M and reliability index, β are defined as:

$$\mu_M = (\mu_C - \mu_D) = (938 - 500) = 438 \text{ kN}$$

β=438/(81^2+200^2)$^{0.5}$=2.03

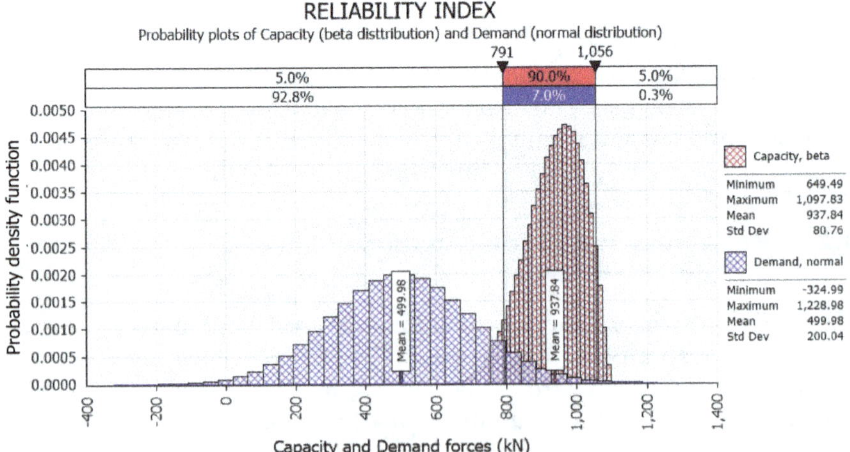

Figure 8.3 Probability plots for Capacity – beta, and Demand – normal distribution (plots generated by @Risk software, Lumivero Inc.)

In Figure 8.3, the Capacity and Demand curves overlap in the central area, showing that a probability of failure exists for conditions where demand exceeds capacity. The probability of failure is proportional to the area of the overlap. That is, the less the uncertainty in the analysis parameters, the narrower the plots and the smaller the area of overlap meaning that the probability of failure is reduced.

8.2.4 Probability distribution for reliability index

The calculated values for μ_M and β listed in Section 8.2.3 above do not account for uncertainty in the values for the Capacity and Demand as shown by the probability distributions in Figure 8.3.

A probability distribution for the Reliability Index can be calculated from the information shown in Figure 8.3 and equation (8.6) using Monte Carlo simulation as follows. In equation (8.6), the numerator $M=(\mu_C - \mu_D)$ is the margin of safety that has a probability distribution that represents the overlap of the Capacity and Demand curves in Figure 8.3, and is equivalent to the probability of failure. Also, in equation (8.3), the denominator $[(\sigma_C^2 + \sigma_D^2)^{0.5}]$ is a fixed value defined by the value of the standard deviation for each probability distribution; for this case, the denominator equals $[(81^2+200^2)^{0.5}=216$ kN].

Monte Carlo simulation involves making multiple calculations of β from equation (8.6), with each calculation using values for μ_C and μ_D being selected, by a random number generator, from the probability distributions. The results of the Monte Carlo simulation to calculate the probability distribution for β is shown in Figure 8.4 where the mean value is $\beta_{mean}=0.98$. This mean value calculated from the probability distributions compares with the deterministic value calculated from the static mean and standard deviations of $\beta=2.03$ shown above – the difference between the deterministic and probabilistic values demonstrates that the influence of uncertainty on design values is to reduce the reliability.

The plot in Figure 8.4 also shows that β has a range of values, defined by maximum and minimum values of the means and standard deviations, consistent with the ranges of the Capacity and Demand distributions shown in Figure 8.3. Also, the β distribution extends to less than 0.0 showing that a probability of failure exists consistent with the overlapping Capacity and Demand distributions.

Figure 8.4 shows that 16% of the Reliability Index distribution is less than 0.0 indicating that the probability of failure for the two overlapping distributions on Figure 8.3 is 16%. This value for $PF=16\%$ for a mean Reliability Index of 0.98 can be compared with the relationship between β and PF shown in Table 8.1.

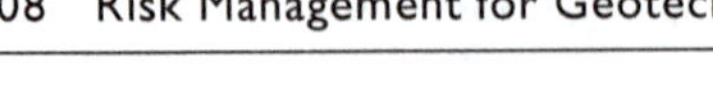

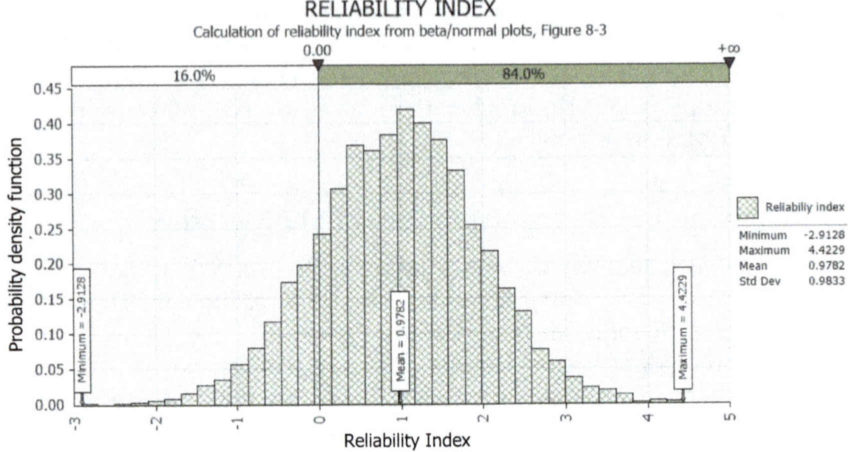

Figure 8.4 Distribution of reliability index calculated from probability distributions of Capacity and Demand shown in Figure 8.3.

8.3 RELIABILITY-BASED DESIGN (RBD)

The previous section of this chapter (Section 8.2) described how the Reliability Index β is calculated from the mean and standard deviation of the factor of safety probability distribution – see equation (8.4) and Figures 8.1 and 8.2. Once the value for β has been established, the probability of failure can be calculated from equation (8.3). This might be described as a "reactive" process in which β is calculated from the probability parameters defining the factor of safety.

An alternative, "proactive" design method is to select a required value for the reliability index that is then used to prepare the design – this procedure is termed Reliability Based Design (RBD). Because the reliability index is directly related to the probability of failure (see equation 8.3), the RBD method allows designs to be prepared with a target probability of failure suitable for the project circumstances, such as $\beta \geq 3$ for a structure with a high consequence of failure, and $\beta \approx 2$ for less critical structures.

Features of RBD are first, each of the parameters for which uncertainty exists can be defined as a separate probability distribution, such as Beta, triangular and normal. In addition, the correlation between parameters can be specified. This allows flexibility in how each parameter is defined, to suit the data that is available. Second, the results of the RBD calculations can be represented in a diagram that clearly shows the relationship between the design point of the structure, and the reliability of the design (Figure 8.6).

Details of the calculation procedures for RBD are beyond the scope of this book, but are described in the book **Reliability-Based Design in Rock**

and Soil Engineering by Dr. B. K. Low published by CRC Press. Associated with Dr. Low's book is a website: www.routledge.com/9780367631390 that contains spreadsheets that perform the RBD design calculations described in the book. Readers with interest in RBD are encouraged to consult Dr. Low's book (Low, 2022).

8.3.1 RBD calculations

The RBD method is demonstrated for the design of a strip footing on soil where the required value of the reliability index is at least 3.0, which is equivalent to a probability of failure, $PF=0.13\%$ for normally distributed parameters (see Table 8.1). For this required value of β, the width of the footing that meets this requirement is determined to be 1.2 m. Figure 8.5 shows the strip footing model, while the diagram in Figure 8.6 illustrates the results of the RBD analysis.

The basis of RBD calculations is to determine from the correlated probability distributions of the design parameters, the relationship between the calculated design point of the structure and the point on the limit state surface (LSS) where the structure is just stable. At this point of stability, termed the "most probable point" (MPP) of failure, the ellipse representing the reliability index is just tangent to the limit states surface. The difference

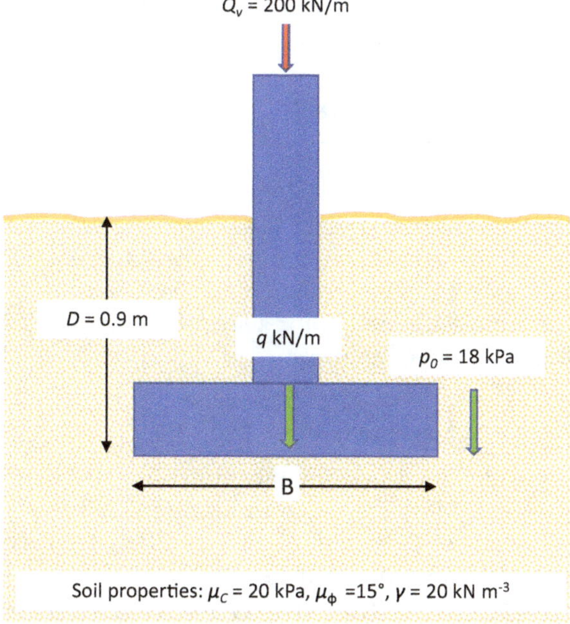

Figure 8.5 Strip footing, embedded to a depth of 0.9 m, with vertical load Q_v bearing on soil used to illustrate RBD method.

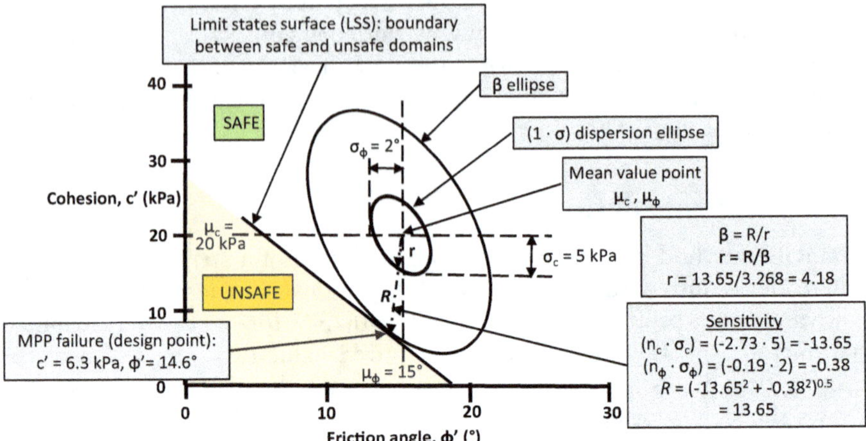

Figure 8.6 Graphical representation of RBD analysis for strip footing illustrated in Figure 8.5.

between the mean value point and the MPP on the LSS plane defines the reliability index, where the LSS plane separates safe combinations of parametric values from unsafe combinations. Figure 8.6 is a graphical representation of the RBD results.

A common calculation procedure for RBD is the First Order Reliability Method (FORM) for correlated non-normals. A special case of FORM is the Hasofer-Lind index for correlated normal random variables (Hasofer & Lind, 1974) that solves equation (8.7):

$$\beta = \sqrt{\left(\frac{x_i - \mu_i^N}{\sigma_i^N}\right)^T \cdot \boldsymbol{R}^{-1} \cdot \left(\frac{x_i - \mu_i^N}{\sigma_i^N}\right)} \tag{8.7}$$

where μ_i^N and σ_i^N are equivalent normal mean and equivalent standard deviation values, and $\boldsymbol{R}$ is the correlation matrix.

The demonstration RBD design for a strip footing discussed below uses the Hasofer-Lind method of analysis.

8.3.2 Performance function

The RBD analysis defines a performance function, $g(x)$ that is identical to the margin of safety, M as discussed in Section 2.2.3 (equation 2.9) and 8.2.4 above. That is, both $g(x)$ and M are the difference between Capacity and Demand, with the structure being stable where $g(x)$ and M are positive, and unstable when they are negative. In RBD analysis, a search is carried out using the FORM procedure to find the conditions where $g(x)=0$. For the example of a strip footing bearing on soil discussed further below, the performance function is:

$g(x)$ = [(ultimate bearing capacity, q_u or Capacity)
– (applied bearing pressure, q or Demand)] (8.8)

where $g(x)=0$ when the search determines the soil strength parameters are reduced to values where $(q_u=q)$.

8.3.3 FORM reliability analysis

The following shows the results of a RBD for the strip footing shown in Figure 8.5 to find the required width of the footing to achieve reliability index of at least 3.0. The calculations were performed with a spreadsheet provided with Dr. Low's book; a printout of the results is shown in Figure 8.7.

The following is a discussion on the elements of the RBD spreadsheet shown in Figure 8.7, with reference to the graphic illustrating the calculation results in Figure 8.6.

- **Equation** $[Q_u = c \cdot N_c + p_0 \cdot N_q + \frac{B}{2} \cdot \gamma \cdot N_\gamma]$ calculates the ultimate bearing capacity of the foundation Q_u as a function of the cohesion c, friction angle ϕ, surcharge pressure of the soil weight at the foundation level p_0, and the unit weight of the soil γ. The calculated values of the three terms in the equation are: $N_q=3.804$; $N_c=10.74$; $N_\gamma=2.507$ as listed on the spreadsheet (Figure 8.7). In the equation for Q_u, the

Reliability-based Design - Strip footing

Calculate required width, B for β ≥ 3

$$q_u = cN_c + p_o N_q + \frac{B}{2}\gamma N_\gamma$$

Units: m, kN, kPa, kN/m³, degrees.

	x*	μ	σ	Corr.Matrix **R**		n
c'	6.339	20	5	1	-0.5	-2.73
ϕ'	14.63	15	2	-0.5	1	-0.19

$$n = \frac{x_i - \mu_i}{\sigma_i}$$

		Q_v	q	B	γ	p_o
N_q	3.804	200	166.7	1.2	20	18
N_c	10.74		Q_v/B			
N_γ	2.507					

g(x) = 0.00

=q_u(x*) - q

$$\beta = \sqrt{\left[\frac{x_i - \mu_i}{\sigma_i}\right]^T [\mathbf{R}]^{-1} \left[\frac{x_i - \mu_i}{\sigma_i}\right]} = 3.268$$

Figure 8.7 Output of spreadsheet used for Reliability Based Design of strip footing shown in Figure 8.5; spreadsheet available at www.routledge.com/9780367631390 (Low, 2022).

N_c term relates to the influence of cohesion, the N_q term relates to the influence of the surcharge, and the N_γ term relates to the influence of the soil weight and foundation width.

- **Shear strength** parameters of the soil, cohesion c' and friction angle ϕ' have mean values μ_c=20 kPa and μ_ϕ=15° respectively, and standard deviations σ_c=5 kPa and σ_ϕ=2°.
- **Corr. Matrix R** defines the correlation between cohesion and friction angle. In this case, the two parameters have a negative correlation coefficient of (0.5) meaning that when the cohesion decreases by 30 kPa, for example, the friction angle increases by 15°.
- **Shear strength values at performance function *g(x)*=0** – solution of the spreadsheet shows that at the MPP point where [$g(x)$=0] on the LSS line, the reduced shear strength parameters are: c'=6.339 kPa and ϕ'=14.63°
- **Parameter n** is sensitivity indicator defined by the equation: [$n=(x-\mu)/\sigma$]. For the condition where the performance function [$g(x)$=0] and the design point lies on the LSS line, the sensitivity indicators are:

c'=6.339 kPa for cohesion, [n_c=(6.339 – 20)/5=–2.73]
ϕ'=14.63° for friction angle, [n_φ=(14.63 – 15)/2=–0.19]

As shown in Figure 8.6, the normalized distances of the mean value point from the design point are:

$(n_c \cdot \sigma_c)=(-2.73 \cdot 5)=-13.65$, and $(n_\varphi \cdot \sigma_\varphi)=(0.19 \cdot 2)=-0.38$

These two values for the normalized distances are used to calculate the value of R:

$R=((-13.65)^2+(-0.38)^2)^{0.5}=13.65.$

For this value for R, and the calculated value for β of 3.268, the dimension of the (1·σ) dispersion ellipse is: [r=(R/β) = (13.65/3.268)=4.18].

- Q_v is the vertical load of 200 kN/m applied to the strip footing.
- **q** is the bearing pressure of the footing with **width B**, on the soil: $q=(Q_v/B)=200/1.2=166.7$ kPa.
- γ is the unit weight of soil at 20 kN/m³.
- p_0 is the soil pressure at the level of the bearing surface, buried to a depth of 0.9 m:

$p_0=(20 \cdot 0.9)=18$ kPa

- β is the reliability index calculated using the Hasofer-Lind index (equation 8.7). For this strip footing example, the search using equation (8.7) found that when the soil shear strength parameters are reduced to **c'=6.339 kPa** and **ϕ'=14.63°**, from the mean values of μ_c=20 kPa and μ_φ=15°, and the footing width B=1.2 m, the bearing pressures are:

[Ultimate bearing pressure (Capacity), q_u=applied pressure (Demand), q=**166.7 kPa**],

and the performance function, $[g(x)=(q_u(x^*)-q=0]$.

where x* is the MPP contact point.

When $[g(x)=0]$, the ellipse representing the reliability index is tangent to the line representing the soil strength and the point of contact (termed the most probable point of failure, MPP) is at shear strength values of $c'=6.339$ kPa and $\phi'=14.63°$.

- **Limit states surface (LSS)** is defined by the shear strength parameters c' and φ' where the foundation is at the point of failure, and the β ellipse just touches the LSS line. The c'/ϕ' values at the contact point (x*) and the mean values μ_c/μ_φ can define quantities that are similar to the partial factors for resistance (φ_g) used in Limit States Design (see Section 2.5 above). That is, for cohesion, the partial resistance factor is [6.33/20=0.3] and for the friction angle, the partial resistance factor is [14.6/15=0.97]. This demonstrates that for RBD, partial resistance factors are calculated rather than selected as is the case for limit states design. These two partial factors can be compared with the recommended values for limit states design listed in Table 2.4 where the resistance factor for cohesion is about 0.6 and the resistance factor for friction is about 0.8. These resistance factors are consistent with the proposition that the resistance factor for cohesion is less than that for friction because cohesion is more difficult to measure and define than friction.

The calculated parameter values are shown graphically in Figure 8.6. This graphic provides intuitive information on the design results. For example, the reliability index will increase if the soil mean value shear strengths were increased by constructing the foundation at a greater depth on a stratum of stronger soil. In this case, the length R, and the dimensions of the β ellipse increase, such that the reliability is greater, and the probability of failure is diminished.

Chapter 9

Professional liability insurance for geotechnical engineering

Most engineering organizations, including individual practitioners, carry Professional Liability Insurance (PLI or Errors and Omissions (E&O) Insurance); this coverage is a requirement of most contracts. Insurance is an essential component of risk management because, however well risk is managed on a project, an element of uncertainty will usually remain, and insurance will limit losses that may be incurred.

Many engineering companies also carry Commercial General Insurance (CGI) that provides coverage for property such as physical equipment, and liability related to bodily injury (see Section 9.4).

9.1 CLAIM HISTORY AND ANALYSIS

The Victor Insurance Company, which offers PLI for many engineering companies in Canada, provides analyses of claims that give a useful insight into the most common causes of claims, and their magnitude (Victor Insurance Managers Inc., 2023). These records likely include all allegations advanced in claims against engineers, including those that have no merit and are subsequently not advanced. While Victor's records are limited to Canada and are for both engineers and architects, it is expected that other countries have comparable values. All insurance companies will keep similar records as part of their business risk management (Figure 9.1).

- **Design errors, 52%** – the most common cause of claims is design errors.
- **Inspection/supervision errors, 14%** – another common cause of claims is errors made by inspectors and supervisors during construction, showing the importance of careful construction inspection. See also Figure 2.7 showing that the annual probability of failure diminishes with increased quality of the engineering employed on projects.
- **Faulty specifications, 2%** – a rare cause of claims is errors in the specifications.

 DOI: 10.1201/9781003271864-9

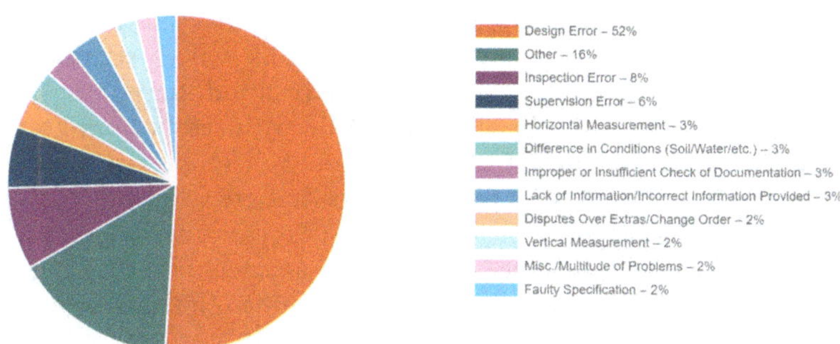

Figure 9.1 Causes of claims in Canada against architectural and engineering companies by allegation (Victor Insurance Managers Inc., 2023).

- **Differing site conditions, 3%** – for all construction claims, only a limited number of claims are related to actual soil and water conditions differing from those expected. However, it is likely that if only projects involving geotechnical engineering were examined, that a higher proportion of the claims would be due to differing soil, rock and water conditions.
- **Miscellaneous causes, 31%** – nearly a third of claims are due to a wide variety of minor issues showing the importance of attention to all details of investigation, design and construction.

This information on the source of claims can be used to focus an organization's risk management efforts in areas that are the most common cause of liability claims, namely design errors.

Further analysis of Victor's records shows that about 80% of claims against engineers and architects are by owners, prime consultants and contractors in the following proportions:

- Private owner – 47% of claims;
- Public sector owner – 17% of claims;
- Prime consultant – 3% of claims;
- Contractors – 12% of claims.

Claims have also been made by subcontractors (1%), adjacent landowners (1%) and injured persons (3%).

9.2 PRINCIPLES OF INSURANCE

The basis of insurance is that risk is transferred from individuals and organizations to insurance companies, and insurance companies manage their risk by insuring a wide variety of risks in terms of both the types of coverage and their location. In this way, if significant damage occurs in an area due to an earthquake or flood, for example, the losses in this area are off-set by the many other insured but undamaged properties that were located in areas subject to different risks.

Insurance is intended to pay for losses that are accidental and may occur in the future. In legal terms, the loss is a "fortuitous" event that is defined as an unforeseen occurrence occurring by accident or chance, not caused by either party, and as such could not be prevented. Furthermore, the insured cannot control the loss. For example, a homeowner cannot burn down his or her own home and collect on a claim. Also, insurance is not provided against a certainty such as wear and tear.

As an alternative to risk transfer, organizations can adopt risk retention or self-insure. That is, governments and large corporations such as railroads have sufficient financial resources to be able to fund their own insurance coverage, although they will need to meet the same regulatory and legal requirements of the insurance industry.

The following is a discussion on the primary components of insurance, with more details of PLI and Commercial General Insurance in Sections 9.3 and 9.4 (Hawrishok, 2019).

9.2.1 Risk

Risk covered by insurance companies is "pure risk" that provides only the potential for financial loss, with no chance of gain or profit. This contrasts with "speculative risk" that provides people with the chance to make either a profit or loss - insurance will not cover speculative risk such as gambling. Furthermore, compensation is up to the amount of the loss and does not provide for profit or financial gain.

Common types of risk that are covered by insurance companies are:

- **Liability risk** is that of being financially responsible for bodily injury or property damage caused by others, or for financial loss resulting from services provided by the insured;
- **Personal risk** – death, illness, disability and unemployment;
- **Property risk** – loss or damage to property owned, rented or leased.

This chapter primarily discusses insurance for professional liability related to engineering services.

9.2.2 Hazard

Hazard is anything that can cause a loss to occur, or increase its severity. The two types of hazards are physical and moral, as follows:

- **Physical hazards** may be related to the location where the engineering services are provided such as in area prone to earthquakes or flooding, and the nature of the business such as design of slopes located in steep, mountainous terrain.
- **Moral hazard** may be related to management issues of engineering companies where, for example, personnel are assigned to work on projects for which they are not properly qualified.

9.2.3 Premiums

The basic financial feature of the insurance industry is that protection against possible losses is provided by the insurance company in exchange for a payment termed a "premium", with a contract (policy) that defines the terms of the protective coverage. Premiums that an insurer receives must be sufficient to pay for losses, expenses, profit and contingencies. Insurance coverage is usually renewable annually, with an adjustment to the terms and the premium, as appropriate.

The premium amount is set by the insurance company and is consistent with the type and magnitude of risk of the coverage that is perceived by the insurance company. That is, coverage of structural engineering services in a high seismic area would probably be charged a higher premium than for a company providing software services. Regarding the relationship between risk and the type of project, projects such as dams and tunnels (heavy civil) will be considered to have a higher risk than buildings (light civil), for example. The premium would also take into account the magnitude of the engineering fees, with the premium being proportional to the anticipated amount of fees. Further details of conditions that are used to determine premium values are discussed in Section 9.3.2.

Insurance companies will keep records of past losses and develop risk profiles of different industries and types of services and set their premium rates accordingly.

9.2.4 Indemnity

Indemnity is the term used for the payment made to the insured to settle claims.

Insurance companies are obligated to pay only such amounts as are required to indemnify people for their losses, and no more. Furthermore, the amount of the insured's actual loss is determined by the value of the loss or damaged property immediately prior to its loss. That is, the loss value is not related to past, or possible future, valuations, but does take into consideration any depreciation that may have occurred since the property was insured. Also, insureds will not be fully indemnified for their loss because the contract will include deductibles, and other clauses that may reduce the amount payable (see Section 9.2.8).

With respect to claims against PLI (professional liability insurance), some valuations may be clear such as the cost of repairs to a building that has been damaged by settlement caused by adjacent construction activities. Other valuations such as costs of delays to project completion, or the required use of more expensive construction methods than those planned by the constructor, may be more difficult to define. For these situations, several parties may be involved such as the owner, the contractor and several engineering companies, each of which will be represented by their own insurance company, and the claim may need to be settled by arbitration or a court case (see Chapter 10). Arbitration or a negotiated settlement is often preferred over a court case because it is quicker, less costly and not public.

If the case involves complex technical issues, it may be necessary for each side to hire expert witnesses with specialist knowledge of the conditions to provide advice to the court; the requirements for expert witnesses are, discussed in Section 10.7. When the value of the damages cannot be clearly defined, a settlement may be negotiated in which the court or the arbitrator(s) take into consideration the evidence presented by all parties. The negotiations can result in several of the parties contributing to the settlement, in proportion to their perceived liability.

9.2.5 Defence costs

If a claim is made against an engineer for a project that is covered by a PLI contract, the insurance company may cover the cost of defending the claim, if that is a condition of the contract. If defense costs are in addition to the amount of the insurance, then the full amount of the coverage is available to settle the claim.

Referring again to Victor Insurance's records (see Section 9.1), it is found that the average cost for defence of a claim is about $C12,000, and that for 63% of claims, the claim is settled without the insurance company paying an indemnity to the claimant. For claims where an indemnity is paid, the average payment is about $C58,000.

9.2.6 Financial reserves

The premiums received by insurance companies are aggregated and invested to develop financial reserves that are then used to pay claims in the future. The insurance industry is highly regulated because their operations are essential to efficient operation of all businesses, and the regulations ensure that they have the financial resources to pay claims promptly, and to be able to cover damages for substantial disasters such as floods or fires. This will require that funds are sufficiently liquid to be readily available to pay claims, and be invested mostly in low risk, long-term financial instruments (bonds) that have limited exposure to adverse market fluctuations.

Once the financial reserve requirements have been met, then excess funds can be returned to the shareholders as dividends.

9.2.7 Reinsurance

Reinsurance is insurance that is purchased by an insurance company (termed the "cedent") from one or more insurance companies as a means of risk management. Reasons for buying reinsurance can be that the reinsurance company will accept risks for higher values than the cedent company wants to insure on it own. Also, reinsurance reduces the potential for a catastrophic loss by the cedent company.

9.2.8 Deductibles

Insurance contracts include deductibles that is the amount the insured pays before the insurance company settles the claim. For example, if an engineering company has PLI with a value of $2 million, the deductible may be between $2,000 and $25,000. The amount of the deductible is selected by the insured, and the amount of the premium decreases if a higher deductible is selected.

Another possible feature of PLI is that the deductible will be reduced by 50% if the claim is settled by arbitration rather than in court.

9.2.9 Adjusters

Adjuster is a term used in the insurance industry for a person who determines the value of a claim. Adjusters may be employees of the insurance company working in the claims department, or are independent contractors. In valuing a claim, adjusters will examine the terms of the insurance contract to make sure that all terms of the contract have been met by the insured.

Adjusters work within their field(s) of expertise, such as commercial buildings, so that they can apply their specialist knowledge to the settlement of the claim.

9.2.10 Contracts

Insurance coverage provided to an engineering company by an insurer will be defined by a contract that is a legal agreement between the two parties that is enforceable by law. Important features of a contract are first, that the parties know the substance and terms of the contract, and that neither party has withheld information that should be known to the other party. Second, the contract must include a consideration that is something of value that has been exchanged between the parties. In the case of a PLI contract, the consideration is the premium paid by the engineering company (insured), and the agreement for the insurer to provide indemnity to the insured in the event of a claim (see Section 10.4.1).

A valid contract can be voided or cancelled by the insurer if an essential element of the contract is missing or compromised. If a contract is voided, then it is assumed that the contract never existed and the entire premium is returned to the insured. If a contract is cancelled, then the insurer must provide coverage during the period that the contract was in effect and can keep the portion of the premium for this period.

9.2.11 Disclosure of material facts

Reasons for cancelling a contract could be failure of the insured to disclose material facts, such as the type of services that the engineering company provides, when signing the contract. Applications for insurance usually include a questionnaire requesting details of the types of projects that have been undertaken in the past year such as heavy civil (e.g., tunnels and dams) and light civil (e.g., highways and railways), and overseas projects, and whether the company is engaged in construction or manufacturing.

The information provided by the insured at the time of signing the contract is usually not verified by the insurer and is accepted in good faith. However, if a claim is submitted, the material facts will be checked, and the contract and coverage could be cancelled if the facts originally submitted were found to be inaccurate.

9.2.12 Claims

Organizations that have an insured loss will normally make a claim to obtain an indemnity (payment) for their loss. When a claim is submitted, the insurer will appoint an adjuster to evaluate the claim because insurers are generally not permitted by law to act as adjusters. When a claim is submitted, the insured is required to provide a statutory declaration (i.e., under oath) as to the cause of the loss, and the amount being claimed.

If an owner of a project, such as a utility company, sues an insured engineering company for damages related to instability of a slope for example, the engineering company will submit a claim to their insurer to be indemnified for the damages related to the slope instability. When the claim involved complex issues, the plaintiff (owner) may engage a law firm to handle the claim, and the insurer for the defendant (engineering company) may engage their own law firm. As the claim develops and if the cause of the loss is found to be poorly defined, the plaintiff's lawyer may decide to sue all engineering companies involved in the project so that each of their actions can be examined in detail; each of these companies would submit claims to their insurers. These examinations may reveal that several engineering companies have responsibility for the slide, in which case, the damages paid by each may depend on the contratural payment terms contained in the policy - see Section 9.2.13, Joint and Several Liability.

9.2.13 Joint and Several Liability

Section 9.2.12 discusses a slope failure that results in damages to a utility company, and the utility company makes a claim to recover the cost of the damages. In this case, the claim is made against the insurance companies that provided coverage for the organizations that are responsible for the design and construction of the slope, which may include the engineer, the contractor and a supplier of materials.

In these circumstances where more than one insurer is liable for damages related to the same claim, the payments made by each insurer may depend, in most common law legal systems, on whether they are "jointly liable", "severally liable" or "joint and severally liable". These terms are defined below.

- **Jointly liable** – if the insurers have joint liability for the damages, then they are each liable up to the full amount of the claim. That is, if one insurer cannot pay their obligation, then the other insurers remain responsible for the full amount of the claim. However, if the full claim is settled by one or two of the insurers, then the utility company cannot make a claim against the insurer that did not contribute to the settlement.
- **Severally liable** – if the insurers are severally liable for the damages, then each insurer is only liable for their respective obligations. In the case of engineering and construction claims, it may be necessary to engage expert witnesses to make assessments of how much responsibility should be assigned to each party. If one party such as the material supplier is found to have no responsibility for the slope failure, then only the insurers for the engineer and the contractor will be required to settle the claim in proportion to their client's level of responsibility for the failure.
- **Joint and severally liable** – under joint and several liability the utility company may pursue claims against any one of the insurers. Joint and several liability is most relevant in tort cases (see Sections 9.3.1 and 10.5) where the utility company may recover all the damages from any one of the insurers regardless of their client's individual share of the liability. One of the consequences of joint and several liability is that

the claim may be made against the party with the greatest financial resources ("deepest pocket") regardless of their client's responsibility. Following this settlement, the insurers may then litigate amongst themselves to better divide liability.

A benefit of joint and several liability is that the probability is maximized that the full damages will be awarded, while the harm is that the recovery may not be made by the party most liable for the damages. As an alternative to joint and several liability, liability may be based on comparative fault in which multiple parties are assigned responsibility for a portion of the damages in relation to the degree of fault that they bear for the damages.

9.3 PROFESSIONAL LIABILITY INSURANCE

Liability risk is, generally, the risk of being held financially responsible for bodily injury or property damage caused to others. With respect to engineering services, PLI contracts protect companies against claims made by clients for inadequate work or negligent actions. PLI policies are also called Errors and Omissions (E&O) insurance.

9.3.1 Tort liability

Settlement of disputes between individuals, corporations and governments, not involving a criminal act, is a matter for civil law, and in particular, tort law. The fundamental purpose of tort is to compensate victims of tort, usually by financial compensation to correct the damages or inadequate performance. Punishment of negligent wrongdoers is not a purpose of tort law.

Because successful tort cases will often require financial compensation to the plaintiffs, it is necessary that engineering organizations have adequate insurance coverage in terms of PLI (PLI) to ensure that funds are available to provide such compensation.

Details of tort law are discussed in Section 10.5 below.

9.3.2 Features of professional liability insurance

The need for insurance coverage for inadequate work or negligent actions is clearly shown by the graphic in Figure 9.1 which shows that 66% of claims against engineers and architects are the result of design error and inspection/supervision error, which may be termed negligence. Insurance protection against claims for negligence is readily available from the insurance industry, but insurance companies practice risk management for their coverage by obtaining detailed information on services provided by the insured. This information allows them to compare these coverage with their records of past claims from which they assess the probability of a claim against the insured during the policy period, and the magnitude of the claim. The assessment of future claims determines the "rate" and the amount of money paid by the insured, termed the "premium", for a specific amount of insurance.

The following is a typical list of information that applicants for insurance provide to insurers from which the premium rate is determined:

i. **Personnel** – qualifications, expertise and years of experience of insured's principals to assess their ability to manage their business risks.
ii. **Gross income** – record of gross income from engineering services to determine the size of the business. Generally, the larger the business, the greater the magnitude of claims being made against the insured.
iii. **Approximate construction values** – this information indicates whether the insured is involved in large projects such as dams and tunnels, or small projects such as residential and commercial buildings. Small projects for which limited engineering is carried out, can be more risky than large, well-engineered projects. For example, Section 2.4 above discusses the relationship between the annual probability of project failure and the level of engineering employed – Best to Poor. That is, it has been found that the annual probability of failure varies from about 1E-7 for well-engineered projects (Best), to about 1E-2 for poorly engineered projects (Poor) (see Figure 2.7). These records indicate that the annual probability of failure of poorly managed projects is about five orders of magnitude greater than for well-managed projects.
iv. **Professional discipline(s)** – disciples such as geotechnical engineering where values of many design parameters may be uncertain, and the consequences of unsatisfactory performance are severe, are likely to be riskier than landscape architecture, for example.
v. **Types of projects** – projects such as dams, tunnels and blasting, where the possible consequences of unsatisfactory performance may be severe, are likely to be more risky than design of building foundations where consequences may be limited to foundation settlement.
vi. **Types of clients** – work for sophisticated clients such, as utility companies and transportation agencies may be less risky than working for small building developers who may limit the scope of site investigations and/or field inspections.
vii. **Concentration of clients** – if more than 25% or 50% of an insured's business is for a single client, this is considered to be a high-risk situation because a claim by this client could negatively affect the insured's business.

viii. **Location of projects** – projects located outside the insured's base area where geological, contractual, personnel and legal conditions may be unfamiliar could be more risky than routine, local projects.
ix. **Subconsultants** – the use of a large number of subconsultants may increase risk because subconsultants may be less engaged in the success of projects compared to the insured's own employees.
x. **Quality control** – it is beneficial to risk management if the insured has procedures to check the quality of designs, and that employees are required to participate in professional development programs (see Section 10.3.1).
xi. **Business activities** – it will be of interest to the insurer if the insured engages in business activities such as construction or manufacturing that are outside their core field of engineering services.
xii. **Claim history** – disclosure of past or active claims against the insured will be relevant to setting the premium rate.
xiii. **Statutory declaration** – the application will require a statutory declaration that the insured will comply with all applicable laws and regulations for their location.
xiv. **Certification** – the application will be signed by a person who has authority to commit the company to the insurance application, and who certifies that the information provided in the application is correct.

9.3.3 Defence costs

PLI contracts may include additional insurance for costs incurred in defending and settling claims, including the costs of defence, costs of obtaining expert witnesses, court costs, and interest on the amount of the judgement. For contracts that include payment of defence costs, defence costs are in addition to the amount of insurance provided by the contract.

The question of coverage for defence costs would need to be addressed for all contracts. For example, defence costs may be covered for claims advanced in one country but not in another.

9.3.4 Amount of insurance

Referring again to data provided by Victor Insurance for PLI policies in Canada, the amount of insurance that is carried by engineering companies depends to some degree on their gross fee income. The amount of insurance ranges from $250,000 for annual fees less than $500,000, to an amount of $10 million for companies with annual fees in excess of $5 million. The most common amount of insurance is $1 million.

It is noted that the contract between an engineering company and a client may stipulate the required amount of insurance. If this amount is greater than the insured's existing coverage, an endorsement, or separate excess professional liability insurance policy with an extra premium, may be required to provide the stipulated coverage.

9.3.5 Claim limits

The amount of insurance that the insured has purchased defines the claim limit that is available. For example, if the amount of the insurance is $2 million, then the per claim limit is $2 million, and the aggregate limit for all claims is also $2 million. The aggregate policy limit applies to the year covered by the policy, and if the policy is renewed the following year, then the new policy's aggregate limit is refreshed and is not impacted by any claims advanced during the previous year.

The difference in the insurance policy between successive years of coverage may be that the premium rate increases for subsequent years, and that the new policy contains specific exclusions.

9.3.6 Contracts

Insurance coverage provided to an engineering company by an insurer, will be defined by a contract that is a legal agreement between the two parties that is enforceable by law. Important features of a contract are first, that the parties know the substance and terms of the contract, and that neither party has withheld information that should be known to the other party. Second, the contract must include a consideration that is something of value that has been exchanged between the parties. In the case of a PLI contract, the consideration is the premium paid by the engineering company (insured), and the agreement for the insurer to provide indemnity to the insured in the event of a claim.

A valid contract can be voided or cancelled by the insurer if an essential element of the contract is missing or compromised. If a contract is voided, then it is assumed that the contract never existed and the entire premium is returned to the insured. If a contract is cancelled, then the insurer must provide coverage during the period that the contract was in effect and can keep the portion of the premium for this period.

9.3.7 Claims-made and occurrence coverage

Two options for liability policies are either claims-made coverage, or occurrence coverage. PLI polices usually have claims-made coverage, while general liability insurance is mostly available as an occurrence policy. With a claims-made policy, coverage is only provided when a claim is filed during the active policy period. In contrast, with occurrence coverage, protection is provided if the loss occurred while the policy was active, but the claim can be filed for a loss after the policy has expired.

The reason for use of claims-made coverage is that it eliminates submission of claims many years after the actual occurrence that is the basis for the allegation. If claims can be submitted after the policy has expired, insurers assume long-term, poorly defined liabilities for which they have to maintain financial reserves to pay possible future claims. For claims submitted long after the project was completed it may be difficult to obtain accurate facts about the loss.

Premiums for claims-made policies are generally lower than premiums for occurrence coverage.

9.3.8 Exclusions from PLI policies

PLI policies will define certain risks for which coverage will not be provided, i.e., exclusions for insurance. Universal exclusions are for events that have the potential to be so catastrophic that insurers cannot afford to include coverage for these events in their standard liability policies. Businesses having exposure to loss from these sources need to make special liability insurance requirements. The following is a list of common exclusions from PLI policies:

- **Contractural liability exclusion** – liability of others assumed under contract, excepting liability that would have exisited in the absence of the contract.
- **Nuclear energy hazard** – coverage not provided for bodily injury or property damage that is required to be insured under a nuclear energy liability policy.
- **War risks** – any bodily injury of property damage arising directly or indirectly out of activities associated with invasion, act of foreign enemy, hostilities (whether war declared or not), civil war and rebellion are not insured by liability policies.
- **Terrorism** – terrorism acts are not covered by the liability policies. The September 2001 attack on the World Trade Centre in New York is estimated to have cost the insurance industry about $40 billion.
- **Pollution, asbestos, fungi and spores** – these hazards have long-term effects that may result in substantial future claims. Pollution liability coverage is available from specialty insurers for businesses that have the potential to cause pollution.

9.4 COMMERCIAL GENERAL LIABILITY INSURANCE

In addition to Professional Liability Insurance (PLI), many engineering companies also carry Commercial General Liability insurance (CGI). Commercial general liability covers physical risks, such as third-party bodily injuries where a customer is injured on company premises, and for property loss and damage. In contrast, PLI covers losses resulting from errors and omissions (E&O) related to the services provided to a third party.

Although CGI policies provide coverage for a wider range of risks than those for PLI policies, the premium for CGI coverage may be less than that for PLI because the likelihood of occurrence of these risks is low. Furthermore, the consequences of losses covered by CGI are generally less than that of professional liability claims, although substantial CGI claims are possible.

The following is a list of risks that can be covered by CGI policies, that can be customized to suit the requirements of each engineering company.

- **Property** – loss or damage of office equipment such as computers and printers, and field equipment such as survey instruments, and loss of important business papers. Coverage may also be provided for damage to buildings.
- **Bodily injury and property damage** – coverage for injury to persons while on the company premises, and damage to the property.
- **Non-owned automobile liability** – coverage for legal liability arising from damage to third-party automobiles.
- **Crime** – coverage for such actions as employee theft, or losses due to forgery.

9.5 FORCE MAJEURE AND ACTS OF GOD

The colloquial term "Act of God" refers to natural disasters (or other destructive events) which are entirely outside of human control; examples of Acts of God include hurricanes, earthquakes, tornados, and tsunamis. The term Act of God does generally not appear in insurance contracts, and is covered by the more general term *force majeure*.

Force majeure is a type of clause included in contracts to remove liability for unforeseeable and unavoidable catastrophes that interrupt the expected course of events and prevent participants from fulfilling their obligations. These clauses generally cover both natural disasters, and catastrophes created by humans. Possible events that would invoke force majeure are extraordinary events or circumstances beyond the control of the parties, such as a war, strike, riot, crime, or pandemic.

Explicitly excluded from force majeure contract conditions is any event described as an Act of God, which is a separate concept within contract law. In practice, most *force majeure* clauses do not excuse a party's non-performance entirely but only suspends it for the duration of the *force majeure*.

It is possible to obtain insurance against events that would be classified as Acts of God, but the coverage would need to be explicitly defined in the contract. For example, it is possible to obtain an endorsement, with payment of an extra premium, for coverage from earthquake damage on a homeowner's insurance policy.

See Section 10.6.7 for a discussion on the legal definition of force majeure related to breach of contract.

9.6 HOW INSURANCE COMPANIES MANAGE THEIR RISK

Because the basis of insurance is the transfer of risk from individuals and businesses to insurance companies, it is fundamental that insurance companies have sophisticated methods of managing their own risk (Global Risk Management Institute, 2018). Insurer's risk management will involve first, that the risk profile of the businesses and properties they insure is diversified, and risk is quantified based on previous loss and claim records. The second requirement is that their financial reserves are protected, as required by legislation to make certain that they can pay claims promptly, and manage high-consequence but low-probability risk events. Financial risk management methods have been developed by the financial industry such as banks and investment companies and are well suited to the requirements of insurance companies.

The six basic measures that apply to financial risk management are as follows:

- Exposure
- Volatility
- Likelihood
- Consequences
- Time horizon
- Correlation

Each of these measures is discussed below.

9.6.1 Exposure

Exposure provides a measure of the maximum potential damage associated with an occurrence. Generally, the risk increases as the exposure to default or losses increases, especially if the risk is not diversified. That is, if the businesses being insured are mainly in coastal areas at risk from hurricanes, then a severe event could cause significant losses for the insurer.

9.6.2 Volatility

Volatility provides a basic measure that can be applied to risk, with risk generally increasing as volatility increases. An example of volatility increasing risk is that future costs of damage become more uncertain which means that it is difficult for insurance companies to predict the cost of possible future losses. If the costs of future losses are uncertain, then insurers need to set aside more funds to cover possible future losses.

Regarding financial volatility, if insurers need to withdraw funds from their reserves at the time of low return on investments, this may negatively impact their finances.

9.6.3 Likelihood

The likelihood of an occurrence is a key measure in risk management. The ability to determine the mathematical probability of an event occurring is the foundation of insurance and risk management. The term "probability" is used when sufficient data (law of big numbers) is available to calculate the probability of an event occurring in the future, based on records of the occurrences of past events. When limited data on past events is available, and the frequency of future events must be estimated, then the term "likelihood" is used. For example, for insurers operating in mountainous terrain, it is likely that extensive information on past slope failures will be available from which annual probabilities can be calculated. However, for rarer events such as earthquakes, the assessment of likelihood by insurers would need to rely on data from a much wider area that may not be directly applicable to their area of business.

9.6.4 Consequences

Consequences are a measure of the degree to which an occurrence could negatively or positively affect an organization.

The relationship between consequence and probability is critical for risk management in assessing risk, and deciding whether and how to manage it. The level of risk is related to the combination of low and high probability of occurrence, and of minor and major consequences. That is, minor consequence events, such as surficial sloughing of soil slopes that is contained in the ditch at the base of the slope would receive little or no attention from the risk management team. However, the risk of major consequence events such as a landslide would be influenced by the history of landslides in the area in order to estimate the probability of landslide occurrence in the future.

9.6.5 Time horizon

The time horizon for exposure is another basic measure of risk management, with long time horizons generally being more risky than short time horizons. Time horizons can be measured in various ways, depending on the application. For engineering applications, it is obviously riskier, for example, to design and construct a dam with a 100-year design life where future capacity and demand requirements are uncertain (see Section 1.6.3 above regarding Black Swan events), than a bridge for a mine access road where the design life is only 10 years.

For financial applications, diversification in financial investments can help to manage the risks associated with the time horizon of these investments. An insurance company that matches the durations of its assets (investments) with its liabilities (loss reserves) neutralizes the risk associated with the time horizon.

9.6.6 Correlation

Correlation is a measure that should be applied to the management of an organization's overall risk portfolio. If two or more risks are similar, they are usually highly correlated, and the greater the correlation, the greater the risk. For insurance of engineering applications, heavy rainfall may cause flooding that inundates buildings in low-lying areas, debris flows that can destroy bridges and buildings, and river scour that can wash out roads and railways. Insurance coverage for these highly correlated events would be risky, and could be balanced by providing insurance in areas with low precipitation and shallow topography.

Chapter 10

Legal principles underlying risk management

As a result of the uncertainties in projects involving soil and rock excavations as discussed in this book, it is inevitable that project outcomes will sometimes differ from what were expected and were required. The possible consequences of these differences in outcomes may be cost overruns and schedule delays, or more severely, physical damage and injuries. In these circumstances, lawsuits may be brought against engineers with the objective of compensating organizations or persons for damages incurred.

This chapter provides an overview of the legal issues related to possible claims against engineers, specifically for geotechnical projects. An important component of risk management is an understanding of the laws applicable to both the engineering profession and the contracts that govern construction projects. This knowledge will be of assistance to avoid situations that can result in lawsuits and illustrate how to defend against lawsuits. Also discussed are requirements applicable to providing expert witness testimony.

Much of the material in this chapter is excerpted from the book ***Law for Professional Engineers*** **by Donald L. Marston** (Marston, 2019). Mr. Marston's book primarily addresses engineering practice in Canada, but also discusses liability for international projects. This chapter references Canadian case law (common-law) which can be considered by courts in other commonwealth jurisdictions and may be persuasive in that court's analysis, but are not binding upon those courts. Mr. Marston book cites many court judgements that have established common-law precedents for a variety of Canadian cases in the construction industry. Readers should consult Mr. Marston's book for more detail on the legal issues discussed in this chapter.

10.1 LEGAL SYSTEMS

In contrast to universally applicable engineering laws, a number of different legal systems are used by countries around the world and the local legal system is applicable to any legal action brought against an engineer in that

DOI: 10.1201/9781003271864-10

country, subject to a contractual term specifying the laws of an alternate jurisdiction. The parties could agree to have the law of a different jurisdiction from where the project is located apply.

Two widely used legal systems are common-law and Napoleonic civil code as discussed below.

10.1.1 Common-law legal system

The most widely used legal system is common-law, which first evolved in the United Kingdom, and is now used in the United States (except in Louisiana which uses a civil law legal system), and most current or former Commonwealth countries such as Australia, Canada (except Quebec) and Singapore.

The following is a brief discussion of the common-law legal system.

The basis of common-law is that, in deciding cases, the courts apply legal principles established in previous court decisions involving similar or analogous fact situations; this is called "the theory of precedent". Departures from established precedent are often slow to evolve. This slow evolution is a characteristic of the common-law legal system where the theory of precedent is of major importance and the basis of predictability in the legal system (Marston, 2019). The courts have also developed a system of equitable relief that is concurrent with the common-law. The ability for courts to dispense equitable relief allows for flexibility in the court's decision-making process so that decisions can reflect society's current values.

The record of legal judgements, including those that have established legal precedents, is available to the public on easily accessible websites. Due to the prevalence of these publicly accessible websites, most courts require counsel to cite "neutral citations" rather than published reporters.

In addition to common-law, also known as judge-made law, an important source of law is "legislation". Legislation consists of statutes enacted by elected legislatures at both the federal level and local levels such as states (in the United States) and provinces (in Canada). Most jurisdictions have specific laws governing the practice of professional engineering and it is important, of course, that individual engineers, and organizations that provide professional engineering services, understand and comply with all relevant legislation. In addition, separate legislation will govern contracts used for construction projects, as well as issues such as copyright and patents. Compliance with these laws is another important component of risk management.

10.1.2 Napoleonic civil code

The other widely used legal system is the Napoleonic civil code, which was established in France in 1804 as a unified set of laws to replace a patchwork of feudal laws that had been in place in France. The Napoleonic civil code, with appropriate local modifications to the code, is now used in countries such as Italy and Spain, and many of their former colonies. Napoleonic civil code is not addressed in this book.

10.2 CIVIL AND CRIMINAL LAW

The two basic divisions of the legal system are civil and criminal law. Civil cases will be the most common for disputes involving geotechnical issues such as changed ground conditions. Criminal cases arising out of construction contracts may be due to fraud or bribery, such as collusion between bidders, or payments to engineers for disclosure to bidders of non-public information.

Basic differences between the criminal and civil law divisions are as follows.

First, in criminal cases, the legal action is between the "accused" and the government[1], In civil cases, the legal action is between the "plaintiff" (e. g, a building owner) and the "defendant" (e. g. an engineering company that provided services to the building owner).

Second, the degree of proof required for a criminal case is higher than for a civil case. In criminal cases, an accused person must be proven guilty "beyond a reasonable doubt". In contrast, in a civil case the plaintiff must prove the case against the defendant by persuading the court, on a "balance of probabilities", that the facts are as the plaintiff alleges them, and that the defendant should be held liable for damages.

10.3 ENGINEERING PROFESSIONAL PRACTICE

Provision of engineering services, with respect to both individual engineers and engineering organizations, is usually closely regulated with penalties for non-compliance. Regulation of engineering is considered necessary because the consequence of failure, or even of unsatisfactory performance, can be significant or even catastrophic particularly for geotechnical engineering projects such as building and bridge foundations, and slopes above important infrastructure.

The following is a summary of the common requirements to practice as a licensed engineer. As the requirements vary among jurisdictions and may be updated and modified from time-to-time, the following discussion should only be used as a guideline.

10.3.1 Engineers

In many jurisdictions, graduate engineers are required to be "licensed" or "registered" if they take responsibility for the preparation of designs

[1] In countries that are part of the commonwealth, criminal cases are brought on behalf of the Crown.

and contracts, inspection of construction work and verification of performance/maintenance. Common terms used for licensed or registered engineers are, for example, Chartered Engineer (CE) in the United Kingdom, Professional Engineer in the United States (PE) and Canada (P.Eng.) and European Engineer (EUR ING) in the countries of the European Union.

The usual requirements for professional registration are a Bachelor's degree in engineering, or possibly a more advanced degree from an accredited university, a stipulated number of years of work and training under the direction of a registered engineer, proficiency in the local language, and possibly passing examinations in subjects such as ethics, for example. In addition, maintenance of an engineer's registration often requires that technical skills be kept up-to-date by a certified annual study program of new developments in the field of practice – these programs may be termed Professional Development Hours (PDH).

Individual engineers may also be required to carry out a risk analysis of their technical competence in relation to the projects they are undertaking. Engineering organizations may need to evaluate the risk level of each project to decide what level of review and checking is required (see Section 9.1 on Claim History and common causes of claims).

10.3.2 Engineering organizations

Organizations providing engineering services will likely be structured as a corporate entity that is either public (owned by public shareholders) or private (owned by its founders, management or a group of private investors) in which the liabilities of the corporation are its own and are not those of its shareholders. That is, the assets of the shareholders are protected from liabilities incurred by the corporation.

Engineering corporations should employ suitably qualified professional engineers, maintain appropriate levels of Professional Liability Insurance (PLI) and Commercial General Liability insurance (CGI) as discussed in Chapter 9, and may be required to register with the local association of Chartered/Professional Engineers and maintain some type of Permit to Practice.

10.4 COMMON CAUSES OF ACTION AGAINST PROFESSIONALS

When considering the narrow legal scope of this book - management of geotechnical risk - the number of reasons that an engineer may be sued for damages is limited, with the two most common reasons being negligence and breach of contract, which are defined as follows:

1. **Negligence** – Most professional liability claims are based on claims of negligence and usually involve an omission or something done negligently, and is a type of tort (see Section 10.5 below). Black's Law Dictionary defines negligence as: "The omission to do something which a reasonable man guided by those ordinary considerations which ordinarily regulate human affairs, would do", in a situation where harm could have been foreseen and prevented (Vaughn, 1977).
2. **Breach of contract** – A breach of contract is a failure, without legal excuse, to perform any promise that forms all or part of the contract. This includes failure to perform in a manner that meets the standards of the industry or the requirements of any express warranty or implied warranty, including the implied warranty of merchantability (Copenhaven, 1992).

Construction claims raise issues that are both technical and legal. The technical issue may be a clear mistake in design, such as the use of an incorrect building code. However, the court will consider legal issues such as who is responsible for the mistake and why, whether anyone relied upon the mistaken design and, if so, what damages occurred as a result and whether there are any clauses in the contract applicable to the claim. Upon consideration of these legal issues, the court may determine that damages should be apportioned between the parties involved in the design and construction process.

The following is a discussion of tort, specifically negligence, and breach of contract.

10.5 TORT LIABILITY

Tort generally refers to a private or civil wrong; negligence is a type of tort. Tort liability may arise independently of contract, such as negligent performance of professional services. It is not necessary for a contract to exist between the engineer and the plaintiff for tort liability to exist (i.e., there is no "privity of contract"). Even services performed gratuitously – without a contract – can give rise to liability in tort if the services are performed negligently.[2]

An example where liability may arise with respect to geotechnical engineering is instability of an excavation in an urban area. That is, an excavated cut face is temporarily supported with tie-back anchors that were installed according to the engineer's design and specifications, but were found to be inadequate resulting in excessive movement of the ground behind the face, causing damage and resulting in additional cost. In this situation, the engineer's liability to the owner might be addressed by the contract between

[2] Liability will only be present if the court determines that the party performing the services owed a duty of care to the person for whom the services were performed. However, the authorities are clear that such a duty may arise where one party has voluntarily undertaken to provide assistance to another but does so negligently – see: *Goodwin v. Goodwin*, 2007 BCCA 81 at para. 26.

them and the damages would relate to the cost of installing additional anchors. However, the situation may develop where the temporary excavation performs satisfactorily, but settlement of an adjacent building occurs, and it is necessary to make repairs to the building. In this case, no contract exists between the engineer and the building owner, but a tort action for negligence can be brought against the engineer by the owner of the adjacent building to recover the costs of the building repairs.

10.5.1 Purpose of tort

The fundamental purpose of tort law is to compensate victims of tort through monetary payments.

Punishment of negligent wrongdoers is not a purpose of tort law. If the circumstances of the tort also constitute criminal activity, then punishment of the criminal would be the subject of criminal prosecution that is independent of civil proceedings.

Negligence claims often require financial compensation to be paid to the plaintiff, either as part of a settlement or an award of damages following trial. Therefore, it is necessary that engineering organizations have adequate insurance coverage in terms of professional liability insurance (PLI) to ensure that funds are available to provide such compensation. PLI will provide protection if an engineer's negligence results in damage; insurance for professional services is discussed in Chapter 9.

10.5.2 Elements of negligence

For a plaintiff to succeed in a negligence claim and satisfy a court that damages should be awarded, the plaintiff must establish that:

a. The defendant owed the plaintiff a duty of care;
b. The defendant breached that duty by his or her conduct;
c. The plaintiff suffered an injury; and
d. The defendant's conduct caused injury to the plaintiff.

If any of these elements of negligence are not established to the legal threshold required, the plaintiff will not succeed.

10.5.3 Engineer's duty of care

The first step in establishing a claim of negligence is to show that the engineer owed the claiming party a duty of care. Typically, an engineer will owe a duty of care to the owner of the project regardless of whether they were retained directly by the owner or through another consultant or contractor. Additionally, it is possible that the engineer will owe a duty of care to other third parties, however, there must be a "proximate" relationship for such a duty to arise.

If the duty of care is novel (i.e. one that has not been previously recognized) then the courts in Canada will perform what is known as the "*Anns/Cooper*" test. This test involves two stages. The first stage is the "proximity" analysis, which involves asking (a) whether the harm that occurred was reasonably foreseeable, and (b) whether there are reasons, notwithstanding proximity, which operate to negate the recognition of a duty between the parties. At this first stage of the analysis, all relevant factors present in the relationship must be examined, including expectations, representations, reliance and the property or other interests involved. Under this step, the fact that the parties could have protected their interests under contract is a crucial consideration. Contractual silence will not automatically foreclose the imposition of a duty of care, but courts must be careful not to disrupt the allocations of risk reflected in the relevant contractual arrangements.

The second stage involves an analysis of the residual policy considerations outside the relationship of the parties that warrant negating the imposition of a duty of care.

10.5.4 Engineer's standard of care

One element in proving a negligence claim is demonstrating a breach of the standard of care required of the defendant. For example, suppose a court is asked to determine if an engineer has been negligent in the performance of engineering services. The court must decide if the engineer's conduct was negligent. The standard applied is based on the premise that engineers have a duty to use the reasonable skill, care and diligence of engineers of ordinary competence. Reasonable skill, care and diligence are measured by applicable professional standards of the engineering profession at the time the services were performed (Marston, 2019), not what may be known later or what may only be seen with the benefit of hindsight.

Regarding the amount of skill required, the engineer need not necessarily exercise an extraordinary degree of skill, unless such a higher standard is specified in the contract. The fact that other engineers of greater experience or ability might have used a greater degree of skill, or even that the engineer might have used a greater skill, is not material to the analysis. Perfection is not the standard, nor required. Failure to conform to standards generally accepted in the profession is strong evidence of negligence.

10.5.5 Disclaimers of responsibility

An issue that may be of importance to the provision of engineering services is the use of engineering reports by third parties. The risk of negligence claims against the original author of the report may be significant if the report is provided, without knowledge of the author, to a third party some years after preparation of the report and in circumstances that differ from those at the time of the report preparation. Examples of differing site conditions that may develop over time, and that invalidate the original report, are changes to the water table or unanticipated slope movement.

A risk management strategy that may be employed by authors of engineering reports for protection from negligence claims by a third party is to include a disclaimer, such as the following:

> ....the report has been prepared by [author] for [client]. The content in the report is based on [author's] best judgement of the information available to it at the time of preparation. Any use that a third party makes of this report, any reliance on, or any decision to be made based on it, are the responsibility of such third parties. [author] accepts no responsibility for damages, if any, suffered by any third party as a result of decisions made, or actions taken, based on this report.

However, there is case law in British Columbia which provides that a disclaimer or limitation of liability clause cannot be introduced after contract formation.[3] Generally speaking, to be enforceable, the term must be brought to the notice of the party at the time the contract was made. If it is not communicated until afterwards, it will be of no effect.

10.5.6 Subsurface conditions

Uncertainty in subsurface conditions frequently exists on geotechnical projects and it is worthwhile to limit the risk that this uncertainty will result in a lawsuit. This uncertainty exists because subsurface investigation methods such as drill holes and test pits can only sample a very small portion of the site's geologic materials, and geophysics, which will provide information on a larger volume of rock or soil, is always open to interpretation. A common method to help quantify uncertainty in subsurface data is to prepare two reports: first, a **Geotechnical Data Report** (**GDR**) that provides only factual information obtained in the investigation without interpretation, and second, a **Geotechnical Baseline Report** (**GBR**) that provides an interpretation of the data across the site that can be used as reference data for all contractors preparing bids for the project (see Section 3.7, Documenting Sub-surface Conditions).

The use of the GDR and GBR approach to provide site information to bidders is commonly used for tunnelling projects where subsurface risk is often significant. Introduction of the practice of preparing GDR and GBR reports has resulted in a substantial reduction in the number of lawsuits related to changed conditions.

Despite evidence showing that disclosure of subsurface information to bidders is beneficial to all parties, owners might include contract wording requiring the contractor to assume all risk for subsurface conditions. An example of a clause attempting to limit the risk of claims for unknown subsurface conditions is as follows:

---

[3] *Repap British Columbia Inc. v. Electronic Technology Systems Inc.*, 2002 BCSC 539.

> "....contractor agrees that under no circumstances whatsoever will it make a claim for reimbursement on account of subsurface conditions which it may claim to encounter during its performance of the work".

Contract clauses such as this may result in contractors including contingency amounts in their bids in an attempt to cover their geologic risk, the consequence of which will be an increase the project price, whether or not adverse conditions were encountered.

A more equitable method of sharing subsurface risk between the owner and contractor is to make payments for actual subsurface conditions that differ substantially from those anticipated, based on the following typical contract wording:

> " ......substantial difference between information related to soil conditions at the work that is contained in the plans and specifications....... supplied to the contractor for its use in preparing the tender..........and the actual soil conditions encountered by the contractor at the work site during performance of the contract".

For such a contract clause to limit the risk of claims, it is necessary, of course, to prepare a detailed report on subsurface conditions that is included in the specifications that the engineer can compare with actual conditions, and quantify the changes. The issue of differences between anticipated and actual soil conditions can be managed using the Difference in Site Conditions (DSC) clause discussed in Section 3.7.4.

Several cases illustrate situations where claims related to subsurface conditions may be successful or unsuccessful, with the decision depending on the terms of the contract. For example, non-disclosure to bidders of important site investigation data can result in a successful claim, or if it is found that conditions differ from information provided in the contract documents. A factor in the success of claims may be whether contractors had an opportunity, or means, to examine the site and make their own assessment of conditions. Examples of favourable site conditions are where, for a rock excavation project, the rock condition can be verified from visible outcrops, and where published information on the location and characteristics of buried utilities proves to be accurate. Less favourable conditions may be where the depth to the groundwater table and the need for pumping to drain the excavation, is unknown, or the soils have very low strength such that substantial support of the excavation is required to prevent settlement of adjacent structures.

Another reason for a successful claim could be where the contract documents specify a required method of construction that is found to be unsuitable for actual site conditions. For example, a claim may be made if trenchless excavation is specified by the owner for installation of an underground utility, but the presence of boulders prevents this method from being used. If an alternative, more costly method is required, the extra costs claimed by the contractor could be the responsibility of the owner.

10.5.7 Limitation periods and discoverability

Limitation statutes in some common-law jurisdictions provide that negligence and breach of contract actions must be commenced within prescribed time periods (Marston, 2019). Should an action be commenced after expiration of the time limit, the action will normally be "statute barred" and may be dismissed on this basis. The prescribed time periods vary with jurisdiction and may be altered with time so engineers should familiarize themselves with local regulations. Limitation periods are usually strictly applied and missing one can have serious legal consequences.

An important issue regarding limitation periods is that negligence and breach of contract claims must be commenced within the prescribed time after the time the "cause of action arose". That is, the limitation period commences when the claim is first discovered, or ought with reasonable diligence to have been discovered, by the plaintiff. This "discoverability" concept, however beneficial to the plaintiff who has suffered the damages, results in the possibility of a claim being made many years after the services have been performed. A possible issue with extended limitation periods is that knowledge of the project eventually diminishes in time and, as such, defence of the claim may be more difficult as project documents are lost and memories fade. As discussed in Section 9.3.7, most professional liability insurance contracts are on a claims-made basis where coverage is only provided when a claim is filed during the active policy period.

10.5.8 Vicarious liability

Courts in Canada have established that an employer is vicariously liable for the negligent performance of its employee. The reasoning behind this concept is that the purpose of tort law is to compensate the injured party, and the employer is better able to provide financial compensation than the individual employee. For this reason, an employer's professional liability insurance should cover the actions of both the company and its employees.

10.5.9 Principles of tort liability

Compensation can be obtained from an engineer both when damage to a person or property has incurred, and when financial loss occurs as the result of advice negligently given – where the person giving the advice knew, or should foreseeably have known, that reliance was being placed on his or her skill and judgement.

Common-law cases in Canada have established the scope and circumstances where engineers have been held responsible for damages (Marston,

2019). These decisions involve a wide range of engineering services, including geotechnical engineering, and provide valuable insight on how engineers can apply risk management principles to limit the risk of being involved in legal action.

The following are some principles that have been established in Canadian common-law regarding tort liability, and how they might apply to a common geotechnical design such as highway construction:

a. **No Contract Required** –for a negligence claim to succeed, it is not necessary that a contract exists between the defendant engineer and a plaintiff such as a contractor or property owner (i.e., there is no "privity of contract"). For example – a road authority contracts with an engineering company to prepare designs and specifications for a road construction project. Following this, the road authority uses the design and specification to contract with a contractor to carry out the construction. During construction, a deficiency in the design is discovered by the contractor who successfully brings a negligence claim against the engineer to recover expenses, despite not having a contract with the engineer.
b. **Reliance on designs** – if the engineer prepares the road design that is subsequently used in construction by a third party such as a contractor, then it is understood that the contractor has to rely on the design for the work. A possible condition in which an engineer's design can no longer be relied upon is where the site geology, such as the rock strength, in some areas of the work, is different from that used in the design, with the result that the planned construction method is not feasible. For example, cost increases would be incurred if the original site information showed weak rock that could be excavated by ripping, but the actual rock strength required the use of blasting that is a more expensive excavation method. For this situation, the issue of liability for extra costs associated with the increased rock strength may depend on whether the different geological condition was known to the engineer at the time of preparing the design, or could have been anticipated based on the investigation program for the project.
c. **Foreseeability** – in preparation of designs, the engineer may make a reasonable attempt to foresee how the project may be used in the future and how such changes in use may invalidate the design. For example, the original road design contract called for a two-lane road. However, the road was subsequently widened to three lanes requiring an increase in the height of an excavated cut. It is found for the new three-lane road that the cut slope angle in the original design is too steep for the increased height, and that slope instability may occur. The engineer's liability for costs related to the slope failures and the need to excavate the slope at a reduced slope angle may depend on whether the design modification to increase the road width was known or contemplated at the time that the original design was prepared.

10.6 BREACH OF CONTRACT

If a party to a contract fails to perform obligations specified in the contract, then the defaulting party has breached the contract. The non-defaulting party is entitled to certain remedies depending on the nature of the breach and the terms of the contract. These remedies could include the non-defaulting party regarding the contract as repudiated and suing for damages sustained because of the breach (Marston, 2019).

An obligation that is essential to a contract is called a "condition", while an obligation that is not essential to a contract is called a "warranty". Breach of either a condition or of a warranty may entitle the non-defaulting party to damages. However, only a breach of a condition that is of fundamental importance to the contract would allow the non-defaulting party to consider the contract discharged by the breach.

Establishing whether an obligation of a contract is a condition or a warranty may be a key issue in a lawsuit. However, it has been determined that a breach of contract is a cause for discharge only if its effect is to render it purposeless for the non-defaulting party to proceed further with performance. To help clarify the circumstances for termination of a contract, construction contracts often contain a special provision stating: "if the engineer determines that the contractor's performance has been inadequate, then the contract may be terminated by the engineer's client, the owner".

10.6.1 Consideration

An essential part of a contract is the *consideration* – the cause, motive, price or compelling influence that induces a contracting party to enter into a contract. For example, an engineering company will enter into a contract with a property owner to design a building for which the engineering company will receive a fee or consideration. However, consideration need not be a monetary payment but can be something of value to which the party is not already entitled, given to the party in exchange for contractual promises. If a contract does not have a consideration, it may be based on a gratuitous promise in which case the party may escape its moral obligation as defined in the contract (Marston, 2019). Contracts may be unenforceable due to a lack of consideration.

10.6.2 Repudiation

When one party to a contract expressly tells the other party that he or she has no intention of performing the contractual obligations, the declaring party has repudiated the contract. If the non-defaulting party treats the contract as discharged, he or she may claim damages against the defaulting party. The right to elect to discharge the contract makes it impossible for the defaulting party to avoid contractual obligations by announcing he or she has no intention of fulfilling the contract.

10.6.3 Remedies

A non-defaulting party is entitled to damages for losses incurred as a result of a breach of contract. The non-defaulting party may also be entitled to equitable remedies such as *quantum meruit*, specific performance or an injunction (Marston, 2019).

The court must determine the amount of damages to be awarded as a result of breach of contract. Damages must flow directly from the breach, or be reasonably foreseeable by both parties at the time of entering into the contract. If the contract was entered into under special circumstances, and if those special circumstances were communicated between the parties at the time the contract was formed, then those special circumstances can be taken into account in determining the damages resulting from the breach.

For a geotechnical project, such contractual special circumstances may relate, for example, to disturbance outside the construction area. For a drill and blast operation, the noise and ground vibration may be sufficient to disturb the neighbours and result in temporary shutdown of the project even if the noise and vibration levels were less than generally acceptable limits, or specified contractual limits. The magnitude of the damages awarded for construction delays may depend on whether the owner knew of the neighbours' sensitivity to the construction activities and notified the contractor.

10.6.4 Direct and indirect damages

In an award for damages, a distinction is generally drawn between direct and indirect damages. Again, referring to blasting for the road construction contract, direct damages could be actual cracking of walls in the nearby houses resulting from ground vibrations, while indirect damages could be reduced productivity and extra cost resulting from the need to detonate smaller than planned blasts, and enforcing restricted working hours.

Indirect damages are sometimes referred to as "special, indirect or consequential" damages. Protection against such indirect damages may be provided by including an appropriate exemption provision in the contract.

10.6.5 Penalty clauses and liquidated damages

Contracts often contain provisions whereby a party is required to pay prescribed damages if a certain event occurs, such as the contract not being completed by a specified date. However, the parties must make a genuine attempt at the time of entering into the contract to pre-estimate the amount of damages likely to occur as a result of such a breach. In contracts, these pre-estimated damages are called "liquidated damages".

The amount of damages awarded by a court will be based on actual damages that result from the breach taking into account the pre-estimation of the damages. In order to be enforceable, liquidated damages must not be a penalty for the breach. Therefore, the word penalty should generally not be used in contracts so as to avoid the connotation of penalizing the contractor.

10.6.6 Quantum meruit

Quantum meruit is a term meaning "as much as reasonably deserved". The court can rely upon this concept as a measure of damages, often in relation to the time spent and materials supplied by the contractor. That is, suppose certain services have been requested and performed, but no express agreement was reached between the parties as to what payment would be provided in return for the services. In such a situation, the court will award payment by implying that the party performing the services ought to be paid a reasonable amount – an amount determined based on *quantum meruit*.

For example, in the case of the road construction contract, if the owner repudiates the contract because it considers that the rock blasting was not satisfactory with respect to noise and ground vibration, then the contractor would consider that the contract is discharged. In this situation, the contractor would attempt to recover outstanding expenses up to the time of the repudiation on a *quantum meruit* basis.

10.6.7 Force majeure

Force majeure is a type of clause included in contracts to remove liability for unforeseeable and unavoidable catastrophes that interrupt the expected course of events and prevent participants from fulfilling their obligations. These clauses generally cover both natural disasters and catastrophes created by humans. Possible events that would invoke force majeure are extraordinary events or circumstances beyond the control of the parties, such as a war, strike, riot, crime, or pandemic. If a force majeure provision is accepted by the contract parties, the contract time is appropriately extended, but the contractor is not entitled to extra payment for costs incurred related to the delay unless the delay was due to the action of the owner or engineer. Precluding payment to the contractor is consistent with the concept that neither party is in a position to control these risks (Marston, 2019)

Explicitly excluded from force majeure conditions is any event described as an Act of God, which is a separate concept within contract law. In practice, most *force majeure* clauses do not excuse a party's non-performance entirely but only suspend it for the duration of the *force majeure*.

Because of the different interpretations of *force majeure* across legal systems, it is common for contracts to include specific definitions of *force majeure*, particularly at the international level. Some systems limit *force majeure* to an Act of God that are natural events beyond human control such as floods, earthquakes and hurricanes, but exclude human or technical failures such as acts of war, terrorist activities, labour disputes, or interruption or failure of electricity or communications systems. When drafting or reviewing a contract, care should be taken to make a distinction between an *act of God* and other types *force majeure.*

See also Section 9.3.8 above regarding exclusions from Professional Liability insurance coverage.

10.6.8 Injunction

An injunction is an equitable remedy, whereby the court commands or prevents an action. The party seeking an injunction must show that the claimant's position has merit, that they will suffer irreparable harm if the injunction is not granted and that, on a balance of convenience, the injunction should be granted.[4] The court has discretion to grant the injunction depending on the particular circumstances of each case and will consider the reasonableness as to the time and conditions to which the injunction applies. A successful claimant may be granted a permanent injunction against a respondent to stop or prohibit him or her from continuing the infringing activity.

Injunctions may be used to enforce a "negative covenant" in a contract. For example, for an excavation contract, the working hours may be limited by local municipal by-laws, and the discharge of groundwater encountered in an excavation may require treatment before it is released into the local storm sewer system. In such circumstances, an injunction could be granted to enforce permitted working hours or prohibit the release of untreated water. However, the courts may apply a test of reasonableness that suits the particular site conditions. For example, allowable work hours may be extended in the event of an emergency, or untreated water may be discharged in the event of flooding that threatens the safety of the site.

10.6.9 Exclusion clauses to limit liability

A means of managing risk in a construction contract is to include exclusion clauses that limit the liability of one of the parties to the contract. For example, exclusion clauses could specify a limited monetary amount for damages should a breach of contract occur, or limit time periods during which legal action may be commenced. However, such exclusion causes are only of value if they are enforceable. A number of cases have considered the enforceability of these clauses.

Canadian courts previously approached the enforceability of exclusion clauses on the "true construction" approach whereby the wording of the exclusion clause is construed in the context of the entire contract. The purpose of this approach is to determine if the parties to the case had intended that the clause would apply.

[4] *RJR Macdonald Inc. v. Canada (Attorney General)*, 1995 CanLII 64 (SCC).

Another approach taken by Canadian courts in determining the enforceability of exclusion clauses is the "doctrine of fundamental breach" of contract. An example of a possible fundamental breach of contract related to a geotechnical project is as follows. The contract includes information on expected subsurface conditions as described in a Geotechnical Data Report (GDR) and a Geotechnical Baseline Report (GBR) (see Section 3.7). The contract contains an exclusion clause stating that the contractor can rely on the described subsurface conditions, and cannot make a claim for extra costs due to site conditions differing from those described in the contract. If actual site conditions are discovered to be materially different, and financially more onerous to the contractor than those described in the GDR and the GBR, a fundamental breach of the contract may be determined. That is, a fundamental breach may occur where the owner does not accept both the validity of this difference in subsurface conditions, and the extra cost incurred by the contractor.

If this action of the owner is determined by the court to be a fundamental breach of the contract, then the exclusion clause is not valid and the contractor may sue the owner for extra costs due to the differing site conditions.

In Canada, court cases since about 2010 have retired the doctrine of fundamental breach of contract as it pertains to exclusion clauses. The courts now apply the following analysis to determine whether an exclusion clause is enforceable.[5]

The first issue is whether, as a matter of interpretation, the exclusion clause applies to the circumstances established in evidence. This will depend on the court's interpretation of the intention of the parties as expressed in the contract. If the exclusion clause applies, the second issue is whether the exclusion clause was unconscionable and thus invalid at the time the contract was made. If the exclusion clause is held to be valid at the time of contract formation and applicable to the facts of the case, a third enquiry may be raised as to whether the court should nevertheless refuse to enforce the exclusion clause because of an overriding public policy. The burden of persuasion lies on the party seeking to avoid enforcement of the clause to demonstrate an abuse of the freedom of contract that outweighs the very strong public interest in their enforcement. Conduct approaching serious criminality or egregious fraud are examples of accepted considerations of public policy that are substantially incontestable and may override the public policy of freedom to contract and disentitle the party from relying upon the exclusion clause.

10.6.10 Equitable estoppel

A general definition of equitable estoppel is that a court will not grant a judgement or other legal relief to a party who has not acted fairly; for example, by having made false representations or concealing material facts from the other party.

---

[5] *Tercon Contractors Ltd. v. British Columbia (Transportation and Highways)*, 2010 SCC 4.

For the doctrine to operate, there must be a legal relationship giving rise to certain rights and duties between the parties; a promise or a representation by one party that they will not enforce against the other their strict legal rights arising out of that relationship; an intention on the part of the former party that the latter will rely on the representation; and such reliance by the latter party. Even if these requirements are satisfied, the operation of the doctrine may be excluded if it is, nevertheless, not "inequitable" for the first party to go back on their promise. The doctrine most commonly applies to promises not to enforce contractual rights, but it also extends to certain other relationships.

In the context of engineering and construction contracts, equitable estoppel can refer to a situation where changes are made to the terms of a contract, and the question arises as to whether these changes were valid and accepted by both parties.

Examples of situations in geotechnical engineering where an equitable estoppel may be applied are as follows:

- For a highway construction project the prime contractor hires a specialist sub-contractor to install piles for a bridge; the contract with the sub-contractor specifies a schedule for the completion of the piling. As the work proceeds the piling work falls behind schedule due to equipment availability issues. The prime contractor accepts the delay but does not document this acceptance. That is, the acceptance of the delay is not accompanied by a consideration, and is therefore not strictly binding. After some time, the pile installation is on the critical path for the bridge construction and the prime contractor requests that the sub-contractor completes the work according to the schedule in the contract. In these circumstances, it may be considered inequitable for the prime contractor to make this demand and it may be "estopped" from enforcing the original contract schedule.
- Another possible equitable estoppel condition may occur in contract administration. For example, a contract requires that contract extras be documented in order that the contractor receive the extra payments. However, if the extras are not documented by the administrator, then the contractor may be equitably estopped (i.e. denied) from its contractual rights to make a claim for extras. This reinforces the importance of recording and documenting the details of construction activities.

10.7 EXPERT WITNESSES

Court cases involving engineering contracts and construction projects will often include technical issues that may not be readily understood by judges and legal counsel representing the parties. In this situation, an engineer may be hired as an expert witness to both analyze the technical issues of the case and to explain the issues to the court. Some of the requirements of an expert witness are as follows:

a. Expert witnesses must be independent of the other parties to the case and have had no prior involvement in the project.
b. Expert witnesses must have specialist knowledge of the technical issues that are the subject of the case and must be recognized by the court as having the expertise to advise the court on these matters.
c. Based on this specialist knowledge, the expert witness may give opinions on the technical issues. In contrast, non-expert witnesses are usually not permitted to give opinions and are restricted to giving factual evidence.
d. The expert witness is often required to produce a report which provides their opinion on the technical issues arising in the case. This report will be circulated to all parties in the case and so must stand up to scrutiny by the opposing side.
e. Expert witnesses are usually engaged by one of the parties, but they should avoid being perceived as allied to their side. To meet this objective, it is important that they provide truthful, objective and independent testimony.
f. Both parties to the case may hire their own expert witnesses and each of them may have different opinions on the facts of the case, and present these opinions in their reports. It is important that the opinions of the other expert witnesses are objectively reviewed and advice given as to their strengths or weaknesses, and are not dismissed as incorrect because they are on the other side of the case.
g. Expert witnesses can expect to be cross-examined by opposing counsel so it is important that they are confident of their facts and can defend their opinion(s). This will require thorough preparation, including consultation with the counsel for their party.

10.8 ARBITRATION AND ALTERNATE DISPUTE RESOLUTION

The preceding topics covered in this chapter are mainly concerned with legal matters related to contracts and construction with the expectation that the disputes will be resolved in court where, in common-law jurisdictions, established precedents will be of great importance. In reality, construction disputes are often settled prior to trial (i.e. "out of court"), but after evidence has been collected and analyzed, often by expert witnesses. If this work shows that the dispute is primarily over technical or monetary issues, a settlement may be negotiated that is acceptable to all parties. A negotiated settlement may be encouraged by insurance companies where they are covering both the defence costs and the settlement, and wish to limit their financial exposure (see Section 9.2.8 regarding deductibles in insurance contracts).

Arbitration and other settlement methods such as mediation and dispute resolution boards are an alternative to litigation that may be less costly and less protracted (although they are not always so). They have the advantage of being private, in contrast to evidence given at trials, which is public. These approaches to settlement of disputes can be termed alternate dispute resolution (ADR) methods.

10.8.1 Arbitration

The use of arbitration to resolve disputes may be specified in the contract as a requirement, or it could be offered as an option. Where arbitration is used, it is usually governed by an arbitration statute, applicable to the local jurisdiction, that sets out the structure and rules of the proceedings. To avoid the risks of settling disputes in unfamiliar jurisdictions, it may be possible to use widely accepted rules established by the 1958 United Nations Convention on the Recognition and Enforcement of Foreign Arbitral Awards ("New York Convention"), to which about 150 countries are signatories.

One of the issues that arbitration statutes may address is how an arbitration decision, if it is disputed by one of the parties, can be appealed to the courts. Generally, the grounds for appealing such a decision are narrow, and related to the conduct of the arbitration rather than the merits of the award.

Arbitrators – an arbitration is conducted by an arbitrator who is independent of the parties in the dispute, but has some knowledge of the technical issues and is familiar with the arbitration process and regulations. Depending on the rules, a single arbitrator may be appointed who is acceptable to both parties, or a three-person panel may be appointed in which each party nominates one arbitrator who then appoints a chair.

10.8.2 Mediation

Mediation is a means of attempting to resolve a dispute through negotiation. In some cases, it can be a requisite preliminary action prior to final and binding arbitration. Mediations usually involve a mediator who is impartial and whose role is to provide guidance to the parties and facilitate the settlement process; the mediator does not act as an arbitrator or judge.

The fundamental difference between mediation and arbitration is that mediation is an informal process by which the parties can reach a mutually agreed settlement. An arbitration is a quasi-judicial adjudication process resulting in a binding decision. In some jurisdictions, mediation and arbitration are separate legal processes.

10.8.3 Dispute resolution boards (DBR)

A DBR's mandate is usually to recommend or decide on solutions to disputes that arise early in the project and is an advisory mechanism to assist parties in resolving disputes as they arise. DRB procedures are intended to be less formal and less time-consuming than arbitration and litigation.

The DRB panel is usually made up of individuals with expertise in the construction or applicable infrastructure industry who are carefully selected based on their neutrality, integrity and expertise.

The DRB panel can make recommendations, or decide on, solutions to disputes, and such recommendations or decisions are contractually admissible as evidence in any subsequent arbitration. However, either party to the contract can object to the decision within a specific time period, in which case the dispute can proceed to arbitration (Marston, 2019).

Appendix I: Glossary of Terms

RISK MANAGEMENT

Annual exceedance probability (AEP): estimated probability that an event of a specified magnitude will be exceeded in a year.

Consequence: in relation to risk analysis, outcome or result of a hazard being realized.

Danger (threat): natural phenomenon that could lead to damage, described in terms of its geometry, mechanical and other characteristics. The danger can be an existing one (such as a creeping slope) or a potential one (such as a rock fall). The characterization of danger or threat does not include any forecasting of occurrence.

Elements at risk: population, buildings and engineering works, infrastructure, environmental features and economic activities in the area affected by a hazard.

Frequency: measure of likelihood expressed as the number of occurrences of an event in a given time or in a given number of trials (see also likelihood and probability).

Hazard: probability that a particular danger (threat) occurs within a given period of time.

Individual risk to life: increment of risk imposed on a particular individual by the existence of a hazard. This increment of risk is an addition to the background risk to life, which the person would live with on a daily basis if the facility did not exist.

Likelihood: conditional probability of an outcome given a set of data, assumptions and information. Also used as a qualitative description of probability and frequency.

Probability: quantitative measure of the degree of certainty. This measure has a value between zero (impossibility) and 1.0 (certainty). It is an estimate of the likelihood of the magnitude of the uncertain quantity, or the likelihood of the occurrence of the uncertain future event.

The two main interpretations for probability are:

i) **Statistical** frequency or fraction that is the outcome of a repetitive experiment such as flipping a coins. It also includes the idea of population variability. Such a number is called an "objective" or relative frequentist probability because it exists in the real world and is in principle measurable by doing the experiment.
ii) **Subjective probability (degree of belief)** quantified measure of belief, judgement, or confidence in the likelihood of an outcome, obtained by considering all available information honestly, fairly, and with a minimum of bias. Subjective probability is affected by the state of understanding of a process, judgement regarding an evaluation, or the quality and quantity of information. It may change over time as the state of knowledge changes.

Risk: measure of the probability and severity of an adverse effect to life, health, property, or the environment. Risk can be also expressed as the product:

Risk = (Probability of an adverse event) · (consequences if the event occurs)

Risk analysis: use of available information to estimate the risk to individuals or populations, property or the environment, from hazards. Risk analyses generally comprise the following five steps:

- Definition of scope, danger (threat) identification,
- Estimation of probability of occurrence to estimate hazard,
- Evaluation of the vulnerability of the element(s) at risk,
- Consequence identification,
- Risk estimation.

Consistent with the common dictionary definition of analysis, viz. "A detailed examination of anything complex made in order to understand its nature or to determine its essential features", risk analysis involves the disaggregation or decomposition of the system and sources of risk into their fundamental parts.

Qualitative risk analysis: analysis which uses word form, descriptive or numeric rating scales to describe the magnitude of potential consequences and the likelihood that those consequences will occur.

Quantitative risk analysis: analysis based on numerical values of the probability, vulnerability and consequences, and resulting in a numerical value of the risk.

Risk assessment: process of making a decision recommendation on whether existing risks are tolerable and present risk control measures are adequate, and if not, whether alternative risk control measures are justified or will be implemented. Risk assessment incorporates the risk analysis and risk evaluation phases.

Risk control: implementation and enforcement of actions to control risk, and the periodic re-evaluation of the effectiveness of these actions.

Risk evaluation: stage at which values and judgement enter the decision process, explicitly or implicitly, by including consideration of the importance of the estimated risks and the associated social, environmental, and economic consequences, in order to identify a range of alternatives for managing the risks.

Risk management: systematic application of management policies, procedures and practices to the tasks of identifying, analyzing, assessing, mitigating and monitoring risk.

Risk mitigation: selective application of appropriate techniques and management principles to reduce either likelihood of an occurrence or its adverse consequences, or both.

Societal risk: risk of widespread or large-scale detriment from the realization of a defined risk, the implication being that the consequence would be on such a scale as to provoke a socio/political response.

Spatial probability: probability that the element at risk is in the area affected by the danger (threat).

Temporal probability: probability that the element at risk is in the area affected by the danger (threat) at the time of its occurrence.

Tolerable risk: risk within a range that is acceptable to society so as to secure certain net benefits. It is a range of risks regarded as non-negligible and needing to be kept under review and reduced further if possible.

Vulnerability: degree of loss to a given element or set of elements within the area affected by a hazard. It is expressed on a scale of 0 (no loss) to 1.0 (total loss).

DECISION ANALYSIS

Branches Lines: lines that connect nodes are called branches. Branches that emanate from a decision node (and towards the right) are called decision branches. Similarly, branches that emanate from a chance node (and towards the right) are called chance branches. In other words, the node that precedes a branch identifies the branch type. A branch can lead to any of the three node types: decision node, chance node, or endpoint.

Chance nodes: circles identify chance nodes; they represent an event that can result in two or more outcomes, such as minor or major damage. Chance nodes may lead to two or more decisions or chance nodes. The sum of all the event probabilities of each node must equal 1.0 showing that these are all the mutually exclusive events that can occur, such as stable slope (P) or unstable slope ($1-P$). Sum of event probabilities: $[P+(1-P)=1.0]$

Decision analysis: formative method for selecting among actions that have uncertain outcomes. This outcome uncertainty can be characterized by probability distributions for variables that represent the key consequences of the considered actions. The decision maker's relative preference for the various possible outcomes can then be described by a utility function that also captures the decision maker's attitude toward risk.

Decision trees: squares identify decision nodes. A decision tree typically begins with a given "first decision" that is called the root node. For example, the root node in a geotechnical design might represent a choice to accept existing conditions, or carry out remedial measures, at a specified cost, to improve stability that will decrease the probability of future instability. The root node is drawn at the left side of the decision tree.

Endpoints: triangles identify endpoints, or termination nodes, which indicate an outcome for that branch. The Endpoint touches one point of the triangle to the branch that it terminates.

Expected Value (EV): value that combines payoffs and probabilities for each event. The greater the EV, the better a particular decision alternative, on an average, when compared to the other alternatives in the decision tree. The EV is calculated for any chance node by summing all the EVs for each branch that is connected to the node. The general formula for calculating EV at any chance nodes is given as:

$$\Sigma\left(EV_{chance\ node}\right) = EV_{branch1} + EV_{branch2} + \ldots + EV_{branchn}$$

where the expected value is given by the product:

Expected Value, EV = (probability of an event occurrence) × (consequence amount, normally a monetary value).

Appendix II: Conversion Factors

Imperial Unit	*SI Unit*	*SI Unit Symbol*	*Conversion Factor (Imperial to SI)*	*Conversion Factor (SI to Imperial)*
Length				
Mile	kilometre	km	1 mile = 1.609 km	1 km = 0.6214 mile
Foot	metre	m	1 ft = 0.3048 m	1 m = 3.2808 ft
	millimetre	mm	1 ft = 304.80 mm	1 mm = 0.003 281 ft
Inch	millimetre	mm	1 in = 25.40 mm	1 mm = 0.039 37 in
Area				
Square mile	square kilometre	km^2	1 $mile^2$ = 2.590 km^2	1 km^2 = 0.3861 $mile^2$
	hectare	ha	1 $mile^2$ = 259.0 ha	1 ha = 0.003 861 $mile^2$
Acre	hectare	ha	1 acre = 0.4047 ha	1 ha = 2.4710 acre
	square metre	m^2	1 acre = 4047 m^2	1 m^2 = 0.000 247 1 acre
Square foot	square metre	m^2	1 ft^2 = 0.092 90 m^2	1 m^2 = 10.7639 ft^2
Square inch	square millimetre	mm^2	1 in^2 = 645.2 mm^2	1 mm^2 = 0.001 550 in^2
Volume				
Cubic yard	cubic metre	m^3	1 yd^3 = 0.7646 m^3	1 m^3 = 1.3080 yd^3
Cubic foot	cubic metre	m^3	1 ft^3 = 0.028 32 m^3	1 m^3 = 35.3147 ft^3
	litre	*l*	1 ft^3 = 28.32 *l*	1 L = 0.035 31 ft^3
Cubic inch	cubic millimetre	mm^3	1 in^3 = 16 387 mm^3	1 mm^3 = 61.024×10^{-6} in^3
	cubic centimetre	cm^3	1 in^3 = 16.387 cm^3	1 cm^3 = 0.061 02 in^3
	litre		1 in^3 = 0.016 39 *l*	1 L = 61.02 in^3
Imperial gallon	cubic metre	m^3	1 gal = 0.004 55 m^3	1 m^3 = 220.0 gal
	litre	*l*	1 gal = 4.546 *l*	1 L = 0.220 gal
Pint	litre	*l*	1 pt = 0.568 *l*	1 L = 1.7598 pt
US gallon	cubic metre	m^3	1 US gal = 0.0038 m^3	1 m^3 = 264.2 US gal
	litre	l	1 US gal = 3.8 *l*	1 L = 0.264 US gal

(Continued)

Imperial Unit	*SI Unit*	*SI Unit Symbol*	*Conversion Factor (Imperial to SI)*	*Conversion Factor (SI to Imperial)*
Mass				
Ton	tonne	t	1 ton = 0.9072 tonne	1 tonne = 1.1023 ton
ton (2000 lb) (US)	kilogram	kg	1 ton = 907.19 kg	1 kg = 0.001 102 ton
ton (2240 lb) (UK)	kilogram	kg	1 ton = 1016.0 kg	1 kg = 0.000 984 ton
Kip	kilogram	kg	1 kip = 453.59 kg	1 kg = 0.002 204 6 kip
Pound	kilogram	kg	1 lb. = 0.4536 kg	1 kg = 2.204 6 lb.
Mass Density				
ton per cubic yard (2000 lb) (US)	kilogram per cubic metre	kg/m^3	1 ton/ yd^3 = 1186.55 kg/m^3	1 kg/m^3 = 0.000 842 8 ton/ yd^3
	tonne per cubic metre	t/m^3	1 ton/ yd^3 = 1.1866 t/ m^3	1 t/m^3 = 0.8428 ton/ yd^3
ton per cubic yard (2240 lb) (UK)	kilogram per cubic metre	kg/ cm^3	1 ton/ yd^3 = 1328.9 kg/m^3	1 kg/m^3 = 0.000 75 ton/ yd^3
pound per cubic foot			1 lb./ ft^3 = 16.02 kg/m^3	1 kg/cm^3 = 0.062 42 lb./ ft^3
	tonne per cubic metre	t/m^3	1 lb./ ft^3 = 0.01602 t/ m^3	1 t/m^3 = 62.42 lb./ ft^3
Pound per cubic inch	gram per cubic centimetre	g/cm^3	1 lb./ in^3 = 27.68 g/ cm^3	1 g/cm^3 = 0.036 13 lb./ in^3
	tonne per cubic metre	t/m^3	1 lb./ in^3 = 27.68 t/m^3	1 t/m^3 = 0.036 13 lb./ in^3
Force				
ton force (2000 lb.) (US)	kilonewton	kN	1 tonf = 8.896 kN	1 kN = 0.1124 tonf (US)
ton force (2240 lb) (UK)			1 tonf = 9.964 KN	1 kN = 0.1004 tonf (UK)
kip force	kilonewton	kN	1 kipf = 4.448 kN	1 kN = 0.2248 kipf
pound force	newton	N	1 lbf = 4.448 N	1 N = 0.2248 lbf
tonf/ft (2000 lb) (US)	kilonewton per metre	kN/m	1 tonf/ft = 29.189 kN/m	1 kN/m = 0.034 26 tonf/ft (US)
tonf/ft (2240 lb) (UK)	kilonewton per metre		1 tonf/ft = 32.68 kN/m	1 kN/m = 0.0306 tonf/ ft (UK)
pound force per foot	newton per metre	N/m	1 lbf/ft = 14.59 N/m	1 N/m = 0.068 52 lbf/ft
Flow Rate				
cubic foot per minute	cubic metre per second	m^3/s	1 ft^3/min = 0.000 471 9 m^3/s	1 m^3/s = 2118.880 ft^3/min
	litre per second	l/s	1 ft^3/min = 0.4719 l/s	1 l/s = 2.1189 ft^3/min
cubic foot per second	cubic metre per second	m^3/s	1 ft^3/s = 0.028 32 m^3/s	1 m^3/s = 35.315 ft^3/s
	litre per second	l/s	1 ft^3/s = 28.32 l/s	1 l/s = 0.035 31 ft^3/s

(Continued)

Imperial Unit	*SI Unit*	*SI Unit Symbol*	*Conversion Factor (Imperial to SI)*	*Conversion Factor (SI to Imperial)*
gallon per minute	litre per second	l/s	1 gal/min = 0.075 77 l/s	1 l/s = 13.2 gal/min
Pressure, Stress				
ton force per square foot (2000 lb) (US)	kilopascal	kPa	1 tonf/ft^2 = 95.76 kPa	1 kPa = 0.01044 ton f/ft^2
ton force per square foot (2240 lb) (UK)	kilopascal	kPa	1 tonf/ft^2 = 107.3 kPa	1 kPa = 0.00932 ton/ft^2
pound force per square foot	pascal	Pa	1 lbf/ft^2 = 47.88 Pa	1 Pa = 0.020 89 lbf/ft^2
	kilopascal	kPa	1 lbft/ft^2 = 0.047 88 kPa	1 kPa = 20.89 lbf/ft^2
pound force per square inch	pascal	Pa	1 lbf/in^2 = 6895 Pa	1 Pa = 0.000 1450 lbf/in^2
	kilopascal	kPa	1 lbf/in^2 = 6.895 kPa	1 kPa = 0.1450 lbf/in^2
Weight Density[a]				
pound force per cubic foot	kilonewton per cubic metre	kN/m^3	1 lbf/ft^3 = 0.157 kN/m^3	1 kN/m^3 = 6.37 lbf/ft^3
Energy				
Foot lbf	joules	J	1 ft lbf = 1.356 J	1 J = 0.7376 ft lbf

[a] Assuming a gravitational acceleration of 9.807 m/s^2.

References

AASHTO. (2015). *LRFD Bridge Design Specifications (7th Edition); Section 10 Foundations*. Washington, DC: American Association of State Highway Transportation Officials.

Ale, B. J., Hartford, D. N., & Slater, D. H. (2021). Prevention, precaution and resilience: are they worth it? *Safety Science 140*(2021), 105271.

Altarejos-Garcia, L., Silva-Tulla, F., Escuder-Bueno, I., & Morales-Torres, A. (2017). Practical risk assessment for embankments, dams and slopes. In K.- K. Phoon, & J. Ching, *Risk and Reliability in Geotechnical Engineering (pp. 437–469)*. Boca Rotan, FL: CRC Press.

Annandale, G. W. (1995). Erodibility. *Journal of Hydraulic Research* 33(4), 471–493.

Annandale, G. W., Abt, S. R., Ruff, J., Whittier, R. (1996). Scour damage. *The Military Engineer 88*(580), 34–35.

Baecher, G. B., & Christian, J. T. (2017). Evolution of geotechnical risk analysis in North American practice. In K.- K. Phoon, & J. Ching, *Risk and Reliability in Geotechnical Engineering (Chapter 12, pp. 471–490)*. Boca Rotan, FL: CRC Press.

Baecher, G. B., & Christian, J. T. (2003). Expert opinion. In G. B. Baecher, & J. T. Christian, *Reliability and Statistics in Geoteechnial Enginering (Chapter 21, pp. 501–522*). Chicester, Sussex, UK: John Wiley & Sons.

Baecher, G. B., & Christian, J. T. (2003). *Reliability and Statistics in Geotechncial Engineering*. Chichester, UK: John Wiley & Sons.

Barton, N. R., Lien, R., & Lunde, J. (1974). Engineering classification of rock masses for the design of tunnel support. *Rock Mechanics* 6(4), 189–239.

Benson, C., Daniel, D., & Boutwell, G. (1999). Field performance of compacted clay liners. *Journal of Geotechnical Engineering 125*(5), 390–403.

Bollaert, E. F., Munodawafa, M. C., & Mazvidza, D. Z. (2013). Kariba Dam Plunge Pool: quasi-3D numerical predictions. *La Houille Blanche 99*(1), 42–49. doi: 10.1051/lhb/2013007.

Bozuzuk, M. (1978). *Bridge Abutments Move (pp. 678)*. Washington, DC: Transportation Research Board, Research Record.

Brummund, W. F. (2011). Geo-risks in the business environment. *Geotechnical Risk Assessment and Management. (pp. 117–128)*. ASCE.

Bunce, C. M., Cruden, D. M., & Morgenstern, N. R. (1997). Assessment of the hazard from rock fall on a highway. Can*adian Geotechnical* Journal 34(3), 344–356.

California Department of Water Resources. (2018). *Independent Forensic Team Report on Oroville Spillway Incident.* Sacramento, CA: California Department of Water Resources.

Canadian Dam Association. (2013). *Guide to Dam Safety Risk Management.* Markham, Ontario: Canadian Dam Association.

Canadian Standards Association. (2019). *Canadian Highway Bridge Design Code (12th Edition).* Ottawa: Canadian Standards Association.Cedergren, H. R. (1989). *Seeepage, Drainage and Flow Nets (3rd Edition).* New York, NY: John Wiley & Sons.

CEN 2004. EN 1997-1:2004. (2004). *Eurocode 7: Geotechnical Design - Part 1: General Rules.* Brussels, Belgium: European Committee for Stadardization.

CEN 2007 EN 1997-2:2007. (2007). *Eurocode 7 Geotechnical Design - Part 2, Ground Investigation and Testing.* Brussels, Belgium: European Committee for Standardization.

CEN, 1991-1. (1994). *Eurocode - Basis of Design and Actions on Structures - Part 1: Basis of Design.* Brussels, Belgium: CEN (European Committee for Standardization).

Chopin, N. (2002). A sequential particle filter method for static models. *Biometrika 89*(3), 539–552.

Christian, J. T., & Beacher, G. B. (2011). *Unresolved Problems in Geotechnical Risk and Reliability. GeoRisk Conference (pp. 50–63).* Atlanta, GA: ASCE.

Christian, J. T., Ladd, C. C., & Baecher, G. B. (1994). Reliabilty applied to slope stability analysis. *Journal of Geotechnical Engineering 120*(12), 2180–2207.

Copenhaven, W. H. (1992). Liability and professional issues facing engineers in industry. *Proceedings IEEE 1992 Annual Textile, Fiber and Film Industry Technical Conference (pp. 201–210).* Charlotte, NC, Inst. Electrical Electronic Engineers.

Cruden, D. M. (1997). *Landslide Risk Assessment.* In D. M. Cruden, Ed. London, UK: Routledge.

Dai, F. C., Lee, C. F., & Ngai, Y. Y. (2002). Landslide risk assessment and management: an overview. *Engineering Geology 64*, 65–87.

Dai, S. H., & Wang, M. O. (1992). *Reliability Analysis in Engineering Applications.* New York, NY: van Nostrand and Reinhold.

Darbourn, K. (2015). *Impact of the Failure of Kariba Dam.* Johannesburg, South Africa: Risk Management Institute of South Africa.

Del Moral, P., Doucet, A., & Jasra, A. (2006). Sequential Monte Carlo samplers. *Journal of Royal Statistical Society: Series B (Statistical Methodology) 68*(3), 411–436.

Duncan, J. M., & Sleep, M. D. (2017). Evaluating reliability in geotechnical engineering. In K.- K. Phoon, & J. Ching, *Risk and Reliability in Geotechnical Engineering (pp. 131–179).* Boca Rotan, FL: CRC Press.

Duzgun, H. S. B, & Lacasse, S. (2005). Vulnerability and acceptable risk in intergrated risk assessment framework. In O. Hungr, R. Fell, R. Couture, & Eberhardt, *Landslide Risk Mangement (pp. 505–515).* Vancouver, Canada: A. A. Balkema.

Essex, R. J. (2007). *Geotechnical Baseline Reports for Construction - Suggested Guidelines.* Washington, DC: ASCE - Technical Committee on Geotehnical Reports of the Underground Technology Research Council.

Fenton, G. A., & Griffiths, D. V. (2008). *Risk Assessment in Geotechnical Engineering.* Hoboken, NJ: John Wiley & Sons.

Fenton, G. A., Naghibi, F., Dundas, D., Bathurst, R. J., & Griffiths, D. V. (2016). Reliability-based geotechnical design in 2014 Canadian Highway Bridge Design Code. *Canadian Geotechnical Journal 53*(2), 236–251.

Ferguson, N. (2021). *Doom - Politics of Catastrophe.* New York, NY: Penguin Books.

Fisher, R. A. (1921). On the mathematical foundation of theoretical statistics. *Philosophical Transactions of the Royal Society, Series A 222*, 309–368.

Foye, K. C., Salgado, R., & Scott, B. (2006). Assessment of variable uncertainties for reliability-based design on foundations. *Journal of Geotechnical and Geoenvironmental Engineering 132*(9), 1197–1207.

Gardner, D. (2009). *Risk.* London, UK: Random House.

Gelman, A. (2004). *Bayesian Data Analysis.* Boca Raton, FL: Chapman and Hall/CRC Press.

Global Risk Management Institute. (2018). *Risk Management Principles and Practices (3rd Edition).* Malvern, PA: The Institutes.

Grimstad, E., & Barton, N. R. (1993). Updating the Q system for NMT (Norwegian Method of Tunnelling). In Kompen, Opsahi, & Berg, *Proceedings of the International Symposium on Sprayed Concrete - Modern Use of Wet Mix Sprayed Concrete for Underground Support.* Oslo: Norwegian Concrete Association.

Guthrie, R. H., & Evans, S. G. (2005). The role of magnitude-frequency relations in regional landslide risk analysis. In O. Hungr, R. Fell, R. Couture, & Eberhardt. *Landslide Risk Management (pp. 375–380).* Vancouver, Canada: Taylor and Francis, London.

Hansen, B. (1956). Limit Design and Safety Factors in Soil Mechanics. Copenhagen, Denmark: Danish Geotechnical Institute, Bulletin No. 1.

Harr, M. E. (1984). *Reliability-Based Design in Civil Engineering.* In Henry M. Shaw Lecture. Raleigh, NC: Department of Civil Engineering, North Carolina State University.

Hasofer, A. M., & Lind, N. C. (1974). Exact and invariant second-order code format. *Journal of Engineering Mechanics, ASCE 110*, 111–121.

Hawrishok, S. (2019). *General and Adjuster Insurance Licensing, Level 1.* Comox, Canada: ILS Learning Corporation.

Hillson, D., & Simon, P. (2020). *Practical Project Risk Management (3rd Edition).* Oakland, CA: Berrett-Koehler Publishers.

Ho, K. K., & Ko, F. W. (2007). Application of quantified risk analysis in landslide risk management practice - Hong Kong Experience. First International Symposium on Geotechnical Safety & Risk (ISGSR 2007). Shanghai, China: Tongji University.

Hoek, E. (1980). *Underground Excavations in Rock.* London, UK: Institute of Mining and Metallurgy.

Hoek, E. (1994). *Acceptable Risk and Practical Decisions in Rock Engineering.* Vancouver, Canada: Tunnelling Association of Canada, 12th Annual Conference.

Hoek, E., & Brown, E. T. (1997). Practical estimates of rock mass strength. *International Journal of Rock Mechanics and Mining Sciences 34*(8), 1165–1186.

Hoek, E., Kaiser, P. K., & Bawden, W. F. (1995). *Support of Underground Excavations in Hard Rock.* Rotterdam: A. A. Balkema.

Hoek, E., & Palmeiri, A. (1998). Geotechnical risk on large civil engineering projects - keynote address. *International Association of Engineering Geologists Congress (pp. 1–11).* Vancouver, Canada: BiTech Publishing.

Hogarth, R. (1975). Cognitive processes and the assessment of subjective probability distributions. *Journal of American Statistical Association 70*(35), 271–294.

ISO. (2019). *Risk Management - Risk Assessement Techniques (IEC 31010, 2nd Edition)*. Geneva, Switzerland: IEC (International Electrotechnical Commission).

Jakob, M., & Friele, P. (2010). Frequency and magnitude of debris flows on Cheekye River, British Columbia. *Geomorphology 114*(3), 382–395.

Jakob, M., & Holm, K. (2012). Risk assessment of debris flows. In J. L. Clague, & D. Stead, *Landslides - Types, Mechanisms and Modelling (pp. 71–82)*. Cambridge, UK: Cambridge University Press.

Jakob, M., Weatherly, H., Bale, S., & Perkins, A. (2017). A multi-faceted debris flood hazard assessment for Cougar Creek, Alberta, Canada. *Hydrology* 4, 7.

Jibson, R. W. (2013). Mass movement causes: earthquakes. In: R. Marston, & M. Stoffel, *Treatise in Geomorphology (pp. 223–229)*. San Diego, CA: Elsevier Inc.

Just case: hazard development at Garibaldi, [1989] 2 SCR 1228 Case Number 20246 (Supreme Court of British Columbia 12 07, 1989).

Kahneman, D., Slovic, P., & Tversky, A. (1982). *Judgement Under Uncertainty: Heuristics and Biases*. Cambridge, UK: Cambridge University Press.

Kalenchuk, K. S., Hutchinson, D. J., & Diederichs, M. S. (2009). Downie slide - interpretations of complex slope mechanics in a massive, slow moving translational landslide. *Proceeding of the Canadian Geotechnical Conference (pp. 367–374)*. Halifax, NS: Department of Geological Sciences and Geological Engineering - Queen's University.

Koelewijn, A. R. (2002). *The Practical Value of Slope Stability Models. Learned and Applied Soil Mechanics Out of Delft* In F. B. J. Barents, & P. M. P. C. Steijger. Lisse, Holland: A. A. Balkema.

Kreyszig, E. (2011). *Advanced Engineering Mathematics (Chapter 25, 10th Edition)*. New York, NY: John Wiley & Sons.

Kulhawy, F. H. (1992). On the evaluation of soil properties. Atlanta, GA, ASCE. Geotechnical Special Publication No. 31 (p. 95–115).

Lacasse, S. (2021). Slope safety - living with landslide risk and climate change. *Slope Stability Seminar*. Santiago, Chile: University of Chile, Department of Civil Engineering. Retrieved from https://www.youtube.com/watch?v=LE3PTTzB1wU&t=3508s

Lacasse, S., & Nadim, F. (1997). *Uncertainties in Characterization of Soil Properties*. Oslo, Norway: Norwegian Geotechnical Institute, Publication No. 201 (pp. 49–75).

Loehr, J. E., Finley, C. A., & Huaco, D. R. (2005). Procedures for Design of Earth Slopes Using LRFD. Final Report R103-030. Columbia, MO: University of Missouri – Columbia.

Low, B. K. (2022). *Reliability-Based Design in Soil and Rock Engineering*. Boca Raton, FL: CRC Press, Imprint of Taylor and Francis Group.

Lumivero Corporation. (2022). *@Risk Probabalitic Risk Analysis software, Version 4.7*. Ithaca, NY.

Macfarlane, D., & Silvester, P. (2019). Performance and management of Cromwell Gorge landslides, Clyde Dam reservoir. *ANCOLD 2019 Conference (pp. 1–19)*. Auckland NZ, Australian National Com. on Large Dams, Hobart, Australia.

Mackay, C. H. (1997). Management of rock slopes on the Canadian Pacific Railway. In D. M. Cruden, *Lanslide Risk Assessment (p. 5)*. London, UK: Routledge.

Marston, D. L. (2019). *Law for Professional Engineers (5th Edition)*. New York, NY: McGraw-Hill Education.

McCann, M. W. (2002). *Fault Tree Analysis - Guide to Risk Analysis for Dam Safety*. Vancouver, Canada: Dam Safety Interest Group, Canadian Electricity Association, II(3.3.2).

Morgenstern, N. R. (1995). Managing risk in geotechnical engineering. *Proceedings of the 10th Pan-American Conference on Soil Mechanics and Foundation Engineering, 3rd Casagrande Lecture (Vol. 4, pp. 102–126)*. Guadalajara, Mexico.

Morgenstern, N. R. (2018). Geotechnical risk, regulation and public policy. *Soils and Rocks 41*(2), 107–129.

Morgenstern, N. R. (2000). Performance in geotechnical practice. *Proceedings Hong Kong, Institute of Engineers Trans. 7*, 2–16.

Nadim, F., Einstein, H., & Roberds, W. (2005). Probabilistic stability analysis for individual slopes in soil and rock. *Landslide Risk Management* (pp. 63–98). Vancouver, Canada: Taylor and Francis.

NCHRP. (2018). Guidelines for Managing Geotechnical Risks for Design-Build Projects (Research Report 884). Washington, DC: National Cooperative Highway Research Program (Transportation Research Board, TRB).

Neal, R. M. (2001). Annealed importance sampling. *Statistics and Computing 11*(2), 125–139.

New Civil Engineer. (2014, April). *Engineers Race to Stop Scour From Undermining Africa's Kariba Dam (pp. 1–6)*. London, UK: Institute Civil Engineers.

New York Times. (2007, July 11). *Boston Big Dig Ceiling Panels*. Retrieved from https://www.nytimes.com/2007/07/11/us/11bigdig.html.

Orr, T. L. (2015). Managing risk and achieving reliable geotechncal designs using Eurcode 7. In K.- K. Phoon, & J. Ching, *Risk and Reliability in Geotechnical Engineering (Chapter 10, pp. 295–433)*. Boca Rotan, FL: CRC Press, Taylor and Francis Group.

Phoon, K. -K. (2016). Role of reliability calculations in geotechnical design. *Georisk Assessment and Management of Risk for Engineered Systems and Geohazards 11*(1), 1–18.

Phoon, K. -K., & Ching, J. (2015). *Risk and Reliability in Geotechnical Engineering*. Boca Raton, FL: CRC Press, Taylor and Francis Group.

Phoon, K. -K., & Kulhawy, F. H. (1999). Characterization of geotechnical variability. *Canadian Geotechnical Journal*, *36*(4), 612–614.

Pierson, L., Davies, S. A., & Van Vickle, R. (1990). Rock Fall Hazard Rating System, Implementation Manual. Washington, DC: Federal Highway Administration, Technical Report FHWA-OR-EG--90-01.

Porter, M., & Morgenstern, N. (2013). *Landslide Risk Evaluation - Canadian Guidelines and Best Practices Related to Landslides: A National Initiative for Loss Reduction*. Ottawa, Canada: Geological Survey of Canada. Retrieved from https://geoscan.ess.nrcan.gc.ca/

Post Tensioning Institute. (2014). *Post Tensioning Manual, 5th Edition, Report PTI DC35.1-14*. Farmington Hills, MI.

Raiffa, H. (1970). *Decision Analysis - Introductory Lectures on Choices Under Uncertainty*. Reading, MA: Addison-Wesley Publising Co.

Read, J. R., & Stacey, P. F. (2009). *Guidelines for Open Pit Slope Design*. Collingwod, Victoria, Australia; Leiden, the Netherlands: CSIRO Publishing; CRC Press/Balkema.

RoadEX Network. (2021). *Risk Management in Highway Engineering, Geotechnical Risk Management*. Stockholm: Roadex Sweden.

Roberds, W. (2005). *Estimating Temporal and Spacial Variability and Vulnerability. Landslide Risk Management*. In Hungr, Fell, Couture, Eberhardt *(pp. 129–157)*. Vancouver, Canada: Taylor and Francis, London, UK.

RocScience Inc. (2021). ROCPLANE. *Planar Wedge Stability Analysis in Rock*. Toronto, Ontario, Canada.

Shaw, W., & Barry, V. (2013). *Moral Issues in Business*. Boston, MA: Wadsworth Cengage Learning.

Silva, F., Lambe, T. W., & Marr, W. A. (2008). Probability and risk of slope failure. *Journal of Geotechnical and Geoenvironmental Engineering 134*(12), 1691–1699.

Silver, N. (2012). *The Signal and the Noise - Why So Many Predictions Fail*. New York, NY: Penguin Press.

Skermer, N. A. (1984). M Creek Debris Flow Disaster. *Canadian Geotechnical Conference: Canadian Case Histories, Landslides (pp. 187–194)*. Toronto, Canada: BiTEch Publishing.

Smith, A. F., & Gelfand, A. E. (1992). Bayesian statistics without tears; a sampling-resampling persective. *The American Statistician 46*(2), 84–88.

Stanford University. (2011). *Life Safety Consequences of Dam Failures in the U. S. National Performance of Dams Program*. Washington, DC: Homeland Security.

Stata Corp LLC. (2021). Stata/MP 17 Statistical Analysis Software. College Station, TX.

Steffan, O. H., Terbrugge, P. J., Wesseloo, J., & Venter, J. (2006). A risk consequence approach to open pit slope design. *Proceedings of the International Symposium on Stability of Rock Slopes in Open Pit Mining and Civil Engineering (pp. 81–96)*. Cape Town: South African Institute of Mining and Metallurgy, Johannesburg.

Straub, D., & Papaioannou, I. (2014). Bayesian updating with structural reliability methods. *Journal of Engineering Mechanics 141*(3). doi:10.1061/(asce)em.1943-7889.0000839.

Straub, D., & Papaioannou, I. (2017). Bayesian analysis for learning and updating geotechncial parameters and models with measurments. In K. K. Phoon, & J. Ching, *Risk and Reliability in Geotechnicial Engineering (Chapter 5, pp. 221–264)*. Boca Raton, FL: Taylor and Francis.

Terzaghi, K., & Peck, R. P. (1967). *Soil Mechanics in Engineering Practice (2nd Edition)*. New York, NY: Wiley.

Tierny, L. (1994). Markov chains for exploring posterior distributions. *The Annals of Statistics*, *22*(4), 1701–1728.

Transportation Research Board (TRB). (2009). *Implementation Status of Geotechnical Load and Resistence Factor Design in State Departments of Transportation (Circular E-C136)*. Washington, DC: Transportation Research Board (TRB).

U. S. Army Corps of Engineers. (1997). *Introduction to Probability and Reliability Methods for Use in Geotechnical Engineering*. Washington, DC: Report ETL 1110-2-547.

U. S. National Committee on Tunnel Technology. (1984). *Geotechnical Site Investigations for Underground Projects*. Washington, DC: National Acedemy Press.

U. S. Nuclear Regulatory Commission (WASH-1400). (1975). *Reactor Safety Study: An Assessment of Accident Risks in U. S. Commercial Nuclear Power Plants*. Washington, DC.

UK Health and Safety Executive. (2010). Guidance on (ALARP) decisions in control of major accident hazards. Merseyside: Health and Safety Executive. Retrieved from https://www.hse.gov.uk/foi/internalops/hid-circs/permissioning/spc-perm-37/

van Staveren, M. (2011). *Uncertainty and Ground Condtions - A Risk Management Approach*. Abingdon, Oxford, UK: Taylor and Francis.

VanDine, D. F., & Lister, D. R. (1983). Debris torrents - a new natural hazard? *British Columbia Professional Engineer, 34*(12), 9–12.

Vaughn, R. C. (1977). *Legal Aspects of Engineering (3rd Edition)*. Dubuque, Iowa: Kendall/Hunt Publishing Co.

Victor Insurance Managers Inc. (2023). *Risk analysis charts*. Retrieved from Property and Liability Insurance: https://www.victorinsurance.ca

Wong, H. N. (2005). Landslide risk assessment for individual facilities. *Proceedings of the International Conference on Landslide Risk (pp. 237–296)*. Vancouver, Canada: Taylor and Francis.

Wu, T. H., Tang, W. H., Sangrey, D. A., & Baecher, G. B. (1989). Reliability of offshore foundations - state of the art. *Journal of Geotechnical Engineering (ASCE) 115*(2), 157–165.

Wyllie, D. C. (1987). Rock slope inventory system. *Proceedings of the Federal Highway Administration Rock Fall Mitigation Seminar*. Portland, OR: FHWA, Region 10.

Wyllie, D. C. (1999). *Foundations on Rock (2nd Edition)*. London, UK: E & FN Spon (Routledge).

Wyllie, D. C. (2014). *Rock Fall Engineering*. Boca Raton, FL: Taylor and Francis.

Wyllie, D. C. (2018). *Rock Slope Engineering (5th Edition)*. Boca Raton, FL: Taylor & Francis.

Wyllie, D. C., Brummund, W. F., & McCammon, N. R. (1979). *Use of Decision Analysis in Planning Rock Slope Stabilization Programs. Transportation Research Board*. Washington, DC: Federal Highway Administration.

Index

www.ingramcontent.com/pod-product-compliance
Lightning Source LLC
LaVergne TN
LVHW010546110826
845149LV00003B/572

* 9 7 8 1 0 3 2 2 2 2 6 7 7 *